THE MEN of MAMMOTH FOREST

A Hundred-year History of a Sequoia Forest and its People in Tulare County, California

by
FLOYD L. OTTER, Manager,
Mountain Home State Forest,
California

1963

Copyright, 1963
by
Floyd Leslie Otter

First Printing

Library of Congress Catalog Card Number:
63-22148

Lithographed in U.S.A.
EDWARDS BROTHERS, INC.
Ann Arbor, Michigan

J H Hastings
12/30/63

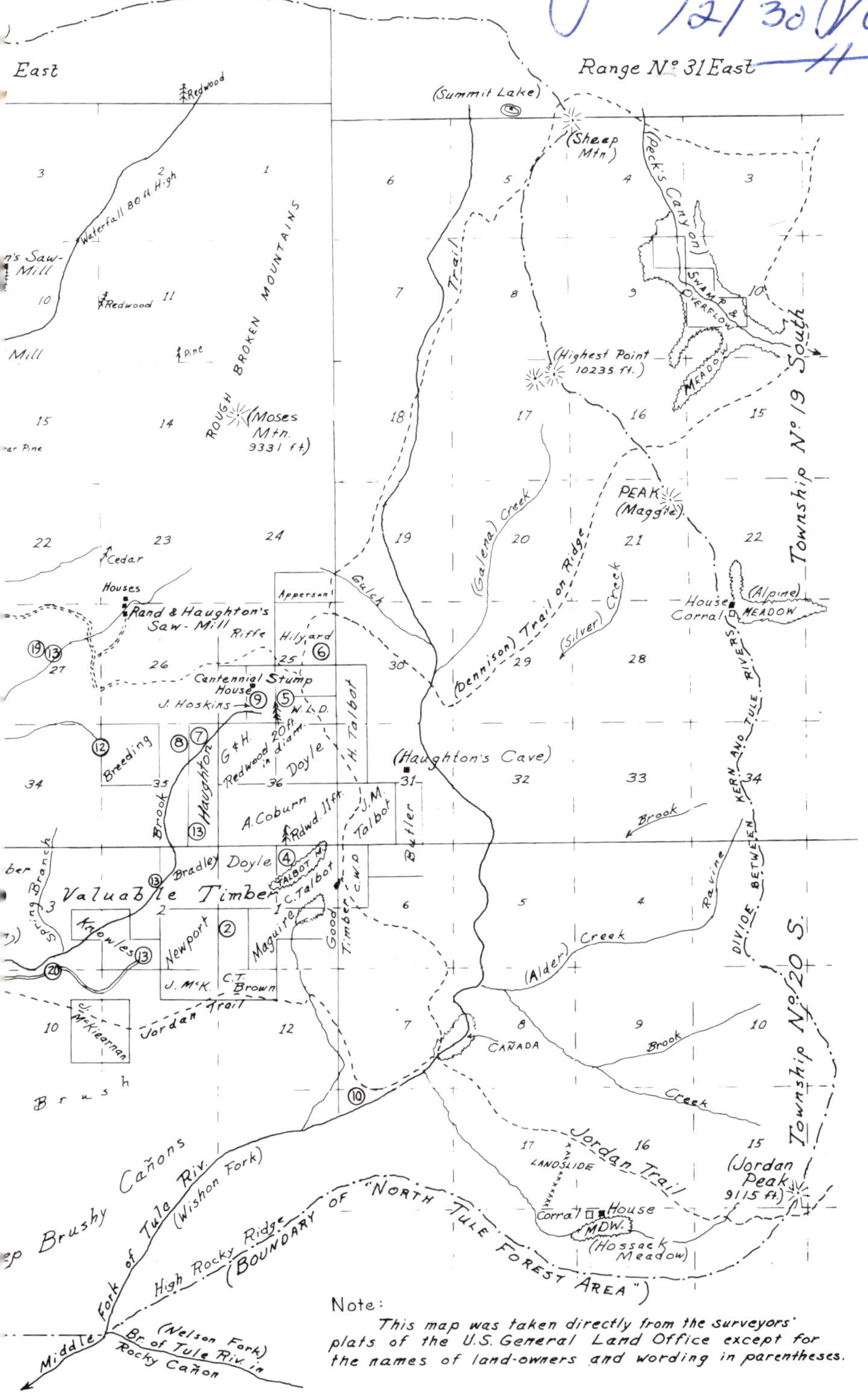

Note:
This map was taken directly from the surveyors' plats of the U.S. General Land Office except for the names of land-owners and wording in parentheses.

To
>	My father,
who, on the high seat of the
grain wagon behind his horses,
told me that land is what holds
the world together.

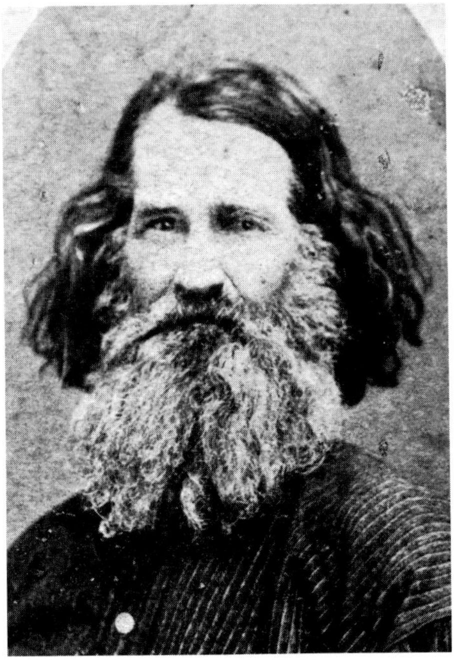

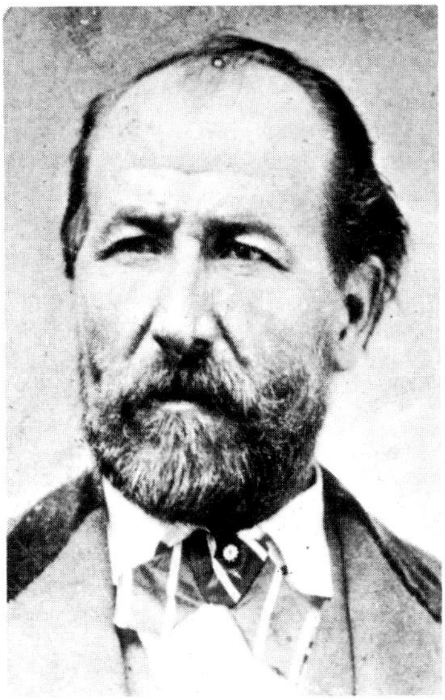

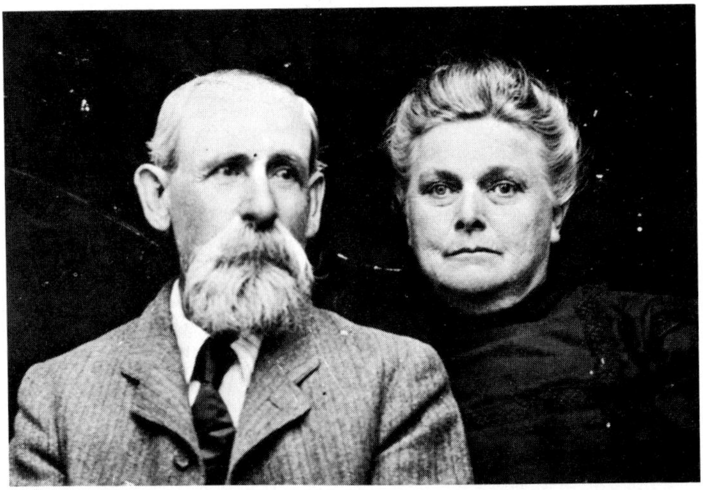

Six of the people who made Mammoth Forest history. Top left to right: timberman Nathan Patrick Dillon when he was a young Visalia businessman; Jacob Cramer, first settler of the North Fork valley of the Tule River; Avon M. Coburn, lumberman and founder of Springville; John J. Doyle, father of Balch Park and dealer in Mountain Home timber and land. Lower left: Andrew Jackson Doty and his wife Sarah, builders and first proprietors of the Mountain Home resort. Photos by courtesy of Mrs. Veda Dillon McCoy, Mrs. Harriett Maxey, Allan Hodge, Mrs. Floreda Marr, and Mesdames Ola Hubbs and Irene Phillips.

A WORD TO THE READER

As the veteran trees in the forest come crashing to earth in a storm, or topple without warning on still summer nights, so, one by one, fall the men and the women who remember the North Tule country in the days of its eager youth. If the story of these people and their mountains is to be told with any resemblance to the way they would tell it, the time has now come.

The only question is, who is to do the telling? Quite obviously, one of these eyewitnesses should have written this book. Among them are individuals who not only remember the old tales, but are endowed with the knack of telling them in a way that would entertain everyone and hurt no one. That they have not more often put pen to paper is our loss.

Why did one such as myself--who claims no special qualifications for the job--undertake to write it? At first I had no such ambition. My only thought was to work my own woodlot, so to speak (which is the Mountain Home State Forest), digging in any spot that showed promise of historical treasure, and chopping the new wood from some old blazes. But it seemed that whenever I went out to work, neighbors would come across their fields or through their woods, and offer to lend a hand. And they would bring treasures of their own that they had laid by on dusty shelves. In this way my neighbors became my friends, interest grew, and facts and artifacts accumulated.

Now our joint harvest is brought together between these two covers. May all who brought their donations find enjoyment in the common hoard, and may others dipping into it make themselves as much at home as life-long residents. The field is not yet exhausted. There is no limit to the historical gems that more diligent digging might unearth. The Mammoth Forest country is a small one and is now relatively uninhabited, but who is to say how many people are required for an Odyssey, or how many acres for an epic.

A hearty "Thank you" is due all of the old-timers and their helpful descendents who searched their memories for dates and their attics for old photographs and papers. Most, if not all, are mentioned in the following pages and thus unavoidably share with the writer any credit or blame that may fall his way.

Some special note must be made of some whose contributions were so basic or so broad that the text and references do them much less than justice. This book would not have been undertaken had not Harold G. Schutt, Editor of <u>Los Tulares</u>, laid the groundwork by his numerous anonymous articles, his enthusiastic searches for old landmarks at Mountain Home, and his encouragement and advice. Invaluable as basic references were Miss Ina H. Stiner's two-volume treasure house of pioneer family histories, genealogies, and photographs filed in the Porterville Public Library. I acknowledge with gratitude the patience of my wife, Mildred, and her assistance and encouragement. Others to whom special appreciation is due include Miss Annie R. Mitchell of Visalia, Secretary of the Tulare County Historical Society; Joseph E. Doctor, Editor of the <u>Exeter Sun</u>; Marion A. Grosse, retired, Fresno State College; Tulare County Librarian Josephine Rhodehamel; Supervisor Eldon Ball, Sequoia National Forest; and C. E. Metcalf, Ray C. Clar, Paul Cox, Lee Burcham, and many others of the California Division of Forestry.

       F. L. O.
       May 1963

CONTENTS

Chapter		Page
	A Word to the Reader	v
I.	The Forest—An Introduction	1
	PART ONE. FIRST ATTACKS ON THE WILDERNESS	5
II.	Far Back on the Trail	6
III.	The Yaudanchi Lose Their Homeland	9
IV.	First Discoveries, Trapping, Mining, and Hunting	20
V.	Trails Across the Sierra	27
VI.	Shepherd's Empire	36
VII.	First Settlers, Sawmills, Roads	46
VIII.	An Old Stump and Two Caves	57
	PART TWO. NOW WE'RE LOGGIN'	65
IX.	A Word to the Wise	66
X.	The Trees Come Down	72
XI.	The Sawmills	83
XII.	Mountain Retreats for the San Joaquin	97
XIII.	Moves Toward Forest Conservation	106
	PART THREE. THE PUBLIC ASSUMES RESPONSIBILITY	111
XIV.	Uncle Sam's Foresters	112

Chapter		Page
XV.	Balch Park	118
XVI.	The State Joins In	123
XVII.	To Sit . . . . To Muse . . . . To Slowly Trace.	141
	REFERENCES	146
	APPENDIX A. ORIGIN OF PLACE NAMES	154
	APPENDIX B. MEMORABLE DATES	159
	INDEX OF PEOPLE AND PLACES	163

CHAPTER I

THE FOREST - AN INTRODUCTION

> Advancing southward the giants become more and more irrepressibly exuberant, heaving their massive crowns into the sky from every ridge and slope, and waving onward in graceful compliance with the complicated topography of the region . . . The extreme upper limit of the belt is reached between the middle and south forks of the Kaweah at an elevation of 8400 feet. But the finest block of the Big Tree forest in the entire belt is on the North Fork of the Tule River. In the northern groves there are comparatively few young trees or saplings. But here for every old storm-stricken giant there are many in all the glory of prime vigor, and for each of these there is a crowd of eager, hopeful young trees and saplings . . . seemingly in hot pursuit of eternal life.
>
> John Muir, *The Mountains of California*

THIS IS A STORY of people and land and trees. It tells how people not greatly different from ourselves--in many cases our own parents or grandparents--found, and took over from a primitive race, a certain small piece of the world; how they used the nation's land laws to bring much of it under their personal ownership; how they attempted to harvest the giant Sequoia trees and other crops produced by a lavish Nature; and how people and events brought about the pattern of land-use and ownership that we know today.

We do not need to say much at this point about these people. They speak for themselves as we go along. We might, with the wisdom of hindsight, condemn them for over-pasturing their "hoofed locusts" in the flowery mountain meadows or for their "rape of the redwoods;" but we are not required to sit in judgment on our forefathers. There is plenty to praise in their courage, fortitude, ingenuity, and persistence, and in their neighborliness and enjoyment of life as they found it. There are even evidences that at times they took on some of the stature of their giant trees.

The trees, too, have a way of unfolding their own story. They were warmed by the same sun as the people, breathed the same air, formed themselves into communities, struggled with one another and with forces beyond their control, and when their time came to go, they too, lay down and made room for another generation.

The land requires a little more introduction. The Mammoth Forest country is as good a land as lays outdoors. True, most of it is steep as a cow's face and some is so precipitous and lonely that even the wary California condor spreads his mighty wings and wheels and soars above it. But the area is not all rocky and on edge. It has its grassy meadows and sunny glades, and most particularly its forests of the tremendous Sequoia gigantea, herein called the Sierra redwood. Some slopes are even so gentle that taking off the timber was, as Springville's lumberman Mal Harris used to say, "like logging a peach orchard."

"Mammoth Forest" is a somewhat fanciful name used here for this area of forested foothills and mountains. Its rains and snows drain into the two northern branches of the Tule River in Tulare County, California. It is a southwestern outpost of the Sierra Nevada, John Muir's "Range of Light." Modern travellers, upon reaching this timbered plateau after the long hot climb out of the haze of the San Joaquin Valley, often sense a feeling of unreality, caused perhaps by the strange brilliance of the sunlight and the sudden change of landscape. The boy in "Jack and the Beanstalk" must have felt this way as he stepped off the top of his magic beanstalk onto the upper-level domain of the giants. The first man to make a written report of his discovery of the Tule River redwoods, Mr. B. W. Farley in 1860, called them "mammoth trees;" and at least one publicity project, Tabor's photographs of 1892, labelled the Mountain Home area the "Mammoth Forest."

Three of the four corners of this area are marked by prominent heights. Brush-covered Lumreau Mountain near the village of Springville is selected as the southwest corner. The Forest Service lookout on Jordan Peak scans the area from the southeast. And craggy Dennison Peak stands sentinel on the northwest. All three are named for men of whom we shall hear later in this narrative. Set like a crown jewel at the fourth and highest corner is Summit Lake near the dividing point of the Kaweah, Kern, and Tule Rivers.

THE FOREST

The acreage of this rumpled landscape is only about 95 square miles, but, as the reader will soon discover, its history involves geography, people, and events of a far wider scope. The area ranges in altitude from about 2000 feet to an unnamed peak 10,235 feet above sea level. Five crooked roads enter the area: the three from the west reach an elevation of about 6500 feet; another from the southwest ends at Camp Wishon, 4500 feet elevation; and the fifth climbs into the southeast corner at Hossack Meadow, where the first planned homesite subdivision of recent times is being developed. The cultivated land in the area is limited to four or five commercial apple orchards and a few home gardens. About twenty year-long residences are located near the western and southern edges. Further back, grouped mainly in two areas, are two or three dozen summer-occupied dwellings of one kind or another. Ownership is about 75 per cent federal (Sequoia National Forest), and 8 per cent state (Mountain Home State Forest). The remainder is in private ownership except for Tulare County's 160-acre Balch Park. Most of the six or seven thousand large redwood trees are on state- and county-owned land.

The highest elevations are wild crags of granite, limestone, and metamorphic rocks interlaced with thickets of wild cherry and bush chinquapin. Groups of western white pine, Jeffrey pine, silver-tipped red fir, and ancient gnarled junipers grow where soil has accumulated. Flocks of band-tailed pigeons find sanctuary here and feast undisturbed on pine nuts. This is the type of country in Tulare County that was the lair of the last grizzly in California. It now provides protection for the common black bear and the rare pine marten and porcupine-eating fisher.

The land between about 3500 and 7000 feet elevation is well covered with virgin and second-growth timber of Sierra redwood, sugar pine, ponderosa pine, white fir, incense cedar, and black oak, with undergrowth of mountain whitethorn, prickly gooseberry, dogwood, and hazelnut. The mule deer, mountain lion, blue grouse, and mountain quail are at home in these forests and the noisy chickaree or Douglas squirrel harvests his unfailing crop of redwood cones.

The foothill country has idyllic, grassy blue-oak groves bright with spring wildflowers and throbbing with bird songs. But it also supports vast tangles of ceanothus, manzanita, chamise, redbud, the showy yellow-flowered flannelbush, mountain mahogany, buckeye, laurel, and other "brush" so dense that at times people and cattle--and perhaps even the deer, coyote, and cottontail--become bewildered.

There have been three periods in the history of this area. The first was a period of small beginnings, given impetus by the so-called Tule Indian War of 1856 and ending with the completion of the public land surveys in 1883. The second period was one of unrestricted and full-blossoming private enterprise, the era of the common man. This expansive period of about twenty years ended as the small land ownerships became consolidated into larger holdings and the newly established forestry branch of the Federal government started active administration of most of the remaining forest land. The third period, from 1905 to 1950, saw the development by two public agencies of forest management on the lands under their jurisdiction and a trend toward recreational use of county- and privately-owned property.

Each period will be discussed separately. Information readily available in printed form is not repeated here. Where evidences of events long past are still recorded on the land or on the trees, special pains are taken to describe these footprints of past generations. Locations of all events are indicated by reference to section numbers of the public land survey. These section numbers, the principal land features, and improvements to 1884 are shown on the map that is a part of this volume.

In telling the story of these people no one is left out just because he was obscure or was a financial failure, and no one is given extra space because he was elsewhere a leading citizen or a "romantic" outlaw. In fact, it is only fair to issue a word of warning to all who have read this far. Although this is a tale of the old West, you will find in it more of standing timber than of bucking horses, more of ringing axes and singing saws than of barking guns, and more of bankruptcies than bank robberies. You meet more in-laws than outlaws, more good cooks than Annie Oakleys; all in all, a breed of men who lived their lives in accord with at least one Biblical precept: "Whatsoever thy hand findeth to do, do it with thy might."

PART ONE

FIRST ATTACKS ON THE WILDERNESS

The Indian's forest: "When I was a little girl I heard several times the story of the Yokuts Land of the Dead, Tih-pik'-nits Pahn. The dead people dance a Lo-ne'-wis, or crying dance there every night. . . . They play games and have a good time. In the morning they all disappear . . . changed into a rotten log, clam shell or something. I never saw them. I just heard Indian doctors say that. But we always go very quiet through a nice place out in the trees."

Yoimut, a Yokuts Indian woman as told to Latta (1949)*.

The sheepman's forest: "In place of a corral, a number of fires are set in fallen timber or living trees, at points which will hem in his flocks for the night to such an extent that wild beasts will be kept off . . . they burn for days, sometimes covering large areas. One can hear the great pines fall in the night. . . ." Elliott (1883).

PRIOR TO THE MIDDLE 1850's all of the ridges, canyons, and meadows of the Tule River headwaters were exclusively Indian country. After the middle 1880's parts of it, including the Mammoth Forest country, were given over almost entirely to the white man's enterprises and pleasures. The intervening thirty years were for him a period of small beginnings in which the wilderness began to give way to attacks of many kinds. The first part of that wilderness to be wiped out was its human component, the Indian, who swiftly met the tragic finish of a century of retreat. Part One describes the struggles and the fate or progress of many of the men of Mammoth Forest of this period, the Indians, the sheepmen, and the first settlers and sawmill men.

---

*Dates in parenthesis refer to publications listed under "References" following Chapter XVII. They are the dates of the publications—not necessarily of the events described.

CHAPTER II

FAR BACK ON THE TRAIL

>Thirty-sixth day - Oct. 26th. This afternoon we broke camp and crossed an oak grove through which flows the San Pedro River, discovered by the expedition of April of this year---1806 . . . a river without water due to its thickness of willows, cottonwoods, "torote" and ash trees, and also because of an abundance of sand. The river from here towards the sierra will be found to have enough water to support a mission. At the sierra it has good water, excellent fields for cultivation, sufficiency of grass, etc. There is a great deal of pine and redwood in the sierra.
>Thirty-seventh day - Oct. 27th. To-day as we travelled upstream we came upon a small rancheria called Coyehete; according to the information gathered from the gentiles it has 400 souls. Fray Pedro Munoz, *Diary* (1806)

FRAY MUNOZ was the diarist for young Gabriel Moraga, an energetic Spanish-Californian Indian fighter and explorer. The above extract provides the first record we have of the discovery of the Tule River and the site of Porterville. The Indian village (rancheria) mentioned was probably within the present city limits. From later evidence it appears that Coyehete was the name of the tribe (Koheti) and that the name of the village was Chokowesho (Latta, 1949).

Apparently the Moraga expedition ascended the Tule no further than the edge of the foothills. Through all the years from 1769 to 1850, when California was being entered, explored, and exploited by men from Spain, Mexico, Russia, Canada, and the United States, the North Tule forest area lay unseen, or at any rate unreported, by white men. Fray Francisco Garces on his long walk north into the San Joaquin Valley in 1776 came only as near as White River. Moraga apparently made two expeditions into the lower Tule River area in 1806. His diarist on four separate occasions mentioned "pine and redwood" but if anyone in the expedition actually saw any Sierra redwood trees, it was in the mountains east of Fresno.* In 1819 an expedition under Jose Maria Estudillo, searching for Indian fugitives from the coastal missions, may have ranged considerably further up the Tule River, judging from an old sketch-map (Cutter, 1950), which pictured a rancheria in the lower Sierra Nevada on the "Rio San Cayetano" (now Deer Creek).

The American mountain men who came into California beginning in 1827 may possibly have seen the Tule River headwaters but their accounts make no mention of it. Jedediah Smith early in 1827 travelled north along the ancient Indian trail that crossed the Tule River near the present Porterville. His trappers found plenty of beaver and otter along the Tule and other San Joaquin streams. During the winter of 1829 and '30 a forty-man party headed by Ewing Young followed Smith's route. They met a group of sixty Hudson's Bay Company men under Peter Skeene Ogden and the two parties trapped beaver and otter together in the San Joaquin Valley. Again in 1833 and the spring of 1834 mountain man Joe Walker circled around the Mammoth Forest country on his trip from Great Salt Lake by a northern route to Santa Barbara and returned via the San Joaquin Valley, Walker Pass, and Owens Lake. He made the first recorded discovery of the giant Sierra redwoods near Yosemite on the northern leg of that journey. John C. Fremont, "the Pathfinder," found Jed Smith's path across the lower Tule River in 1844, but went no further up the river than those who had passed by before. All of these expeditions are described in more or less detail by Smith (1939), Caughey (1940), Cutter (1950), Leader (1928), and Doctor (1959).

It was the year 1850 before there is any definite record that any white man penetrated into the Tule River foothills. Lieutenant George Horatio Derby of the Topographical Engineers, United States Army, camped on Deer Creek near the site of Terra Bella; and on a 40-mile side trip on May 7, 1850 led his eight men up Deer Creek (which he called "Moore's Creek") to somewhere near California Hot Springs, an area "heavily timbered with oaks and three large species of pine" (Derby, 1852), and thence "through the ravines" to an Indian rancheria

---

*Some historians believe Moraga's men mistook the incense cedars for redwoods, but we should give the Spanish and their Indian guides credit for knowing their trees. They lived much closer to Nature than we do and were well acquainted with the Coast redwoods near Monterey.

on the Tule River. This was probably on the South Fork at or near the present community center on the Tule River Indian Reservation.

If, during all this time, any Spanish miner, American trapper, or nondescript renegade or deserter saw the great trees of Mammoth Forest he did not succeed in leaving a record of it. Unfortunately for us, some of the hardy souls who roamed the most wrote the least.

CHAPTER III

THE YAUDANCHI LOSE THEIR HOMELAND

> In nature's infinite book of secrecy
> A little can I read.
>
> Shakespeare, *Anthony and Cleopatra*.

"ON EVERY HILL burned the signal fires of the Indians," says George Creel (1926) in telling the story of Kit Carson's travels through the San Joaquin and Sacramento Valleys in the spring of 1830. Carson was a member of Ewing Young's fur-trapping expedition. Creel continues: "And when it was seen that the savages meant war Carson urged the wisdom of a bold stroke that would instill a wholesome fear. Picking a handful of the best riflemen, he fell on an Indian village in the dead of night, wiping it out. Then, following with the tenacity of a hound, he gave successful battle to the broken remnants of the tribe in a mountain gorge."

We do not know what village Carson wiped out nor to which mountain gorge the broken remnants fled to make their last stand, but the incident shows that the Indians of the Sierra and the central valley did not give up their country without a fight. It is also an example of the one-sided outcome of most of the encounters. In this instance the Indian tribes were probably especially alert to an invasion because only a year earlier the Indian leader Estanislao had organized the northern San Joaquin Valley tribes in a war against the Mexican authorities that was not as one-sided as most.

The signal fires were an important part of the Indians' communication system. Researches by Donald Witt of Porterville and others indicate, for example, that word could be sent in a matter of minutes from certain signal hills near Porterville to the upper reaches of Tule River. The point known as "Snail Head" just south of Springville is supposed to have been one of the main stations. Each Indian chief of the tribes of this area had special assistants called "winatuns" stationed on either side of his village to warn him of approaching danger, greet friendly visitors, and send signals and runners.

The Yaudanchi. The Indians who claimed as their home the North Fork of the Tule River and the main river above Success Dam called themselves Yau-dan'-chi, a word thought to mean "uplanders." Before the white man's diseases (syphilis before 1820, smallpox in 1833, cholera in 1849, and measles in 1887) spread to them from the Spanish coastal settlements and from the various expeditions into the San Joaquin Valley, the Yaudanchi may have numbered as many as the present white population of the same area. Prior to 1800 they had eight villages, according to Cook (1955), scattered along all of the branches of the Tule. A map in Kroeber's Handbook (1953) shows a village named Uk-un'-ui (meaning "drink") just above Springville and another called Shawahtau further up the Middle Fork. There are rock paintings and many other evidences of Indian occupancy at Rancheria, but this was probably only a temporary food-gathering camp. The main Indian villages on the North Fork were probably near the old Milo Post Office and Battle Mountain.

The Yaudanchi tribe was one of approximately fifty independent "nations" that spoke various dialects of a language called "Yokuts" (Heizer and Whipple, 1960). These tribes occupied most of the San Joaquin Valley. Kroeber describes them as follows: "They were a tall, well-built people of open outlook, were the Yokuts: frank, up-standing, casual and unceremonious, optimistic and friendly, fond of laughter, not given to cares of property or too much worry about tomorrow; and they lived in direct simple relation to their land and world, and to its animals, spirits, and gods, and to one another." (Latta, 1949). Latta says he knew one Indian woman who could point out and give the Indian names for more than 500 species of wild plants.

There appears to be no record that any white man, Spaniard or American, saw any of the upper Tule River villages (Yaudanchi and Bokninuwadi) until the visit by Lieutenant Derby, mentioned in Chapter II. Derby (1852) says, "We suddenly came upon a rancheria of Indians in a sequestered nook in the hills. We swam the river and were met on the bank by all the men (60 or 70) belonging to the band. They received us favorably, although with evident distrust. . . . I had been previously

OTHER TRIBES

told by the captain of the Ton Taches that they were a hostile, thieving nation. . . . I suspected that nothing but our number and the well-armed condition of the party prevented our being treated with incivility." Many other Americans undoubtedly visited Yaudanchi villages between 1850 and 1856 and even built cabins in the area, but scant record of their observations or experiences has come down to us.

Other Tribes. The forest areas at the head of the Tule River--our "Mammoth Forest"--were used as food-gathering grounds by several tribes. The Yaudanchi was the only Yokuts tribe that bordered this forest area, but several alien tribes from the north and east also used it, and very probably had a firmer hold on it than the Yokuts. These other tribes spoke dialects classed as Shoshonean. These tribes as a group the whites called Paiute, Mono, Monache and other names, often preceded by uncomplimentary adjectives. For simplicity this whole group, plus any others of the Shoshonean "family" who came into the area from Owens Valley or Nevada, are here lumped together as "Paiutes." The nearest of the Paiute tribes to the Yaudanchi were the Tu'-ba-tu-la'-bal (also called Pi-tanisha) of the upper Kern River, and the Balwisha (Potwishi) of the Kaweah River foothills. All of these various tribes probably used the North Tule forest area for collecting sugar pine nuts and wood for bows and arrows, and for hunting certain game animals. Latta (1949) says they preferred the mountain acorns over those of the foothills. This, together with the cooler climate, was enough to make them frequent the mountains in summer and fall. The Tubatulabal (meaning "pine-nut eaters") were not a typical Paiute group. They were more friendly to the Yokuts and intermarriage was common (Kroeber, 1953). Excursions to Mountain Home, therefore, did not originate with the white people. No doubt many a coy Yaudanchi maid has been wooed and won under the towering redwoods, while not overly watchful grandmothers sat on a sunny rock exchanging gossip and pounding pine nuts for the winter food supply.

Grinding Holes and Rock Basins. Indian grinding holes and other evidences of Indian occupancy are not as common in the mountains of the Tule watershed as they are in the foothills. Typical small grinding holes in granite rock have been found at only ten locations above 5500 feet in elevation in the Mountain Home area, invariably near one or more large "Indian bathtubs." They are also found at about 6000 feet in Dillonwood.

The large prehistoric rock basins (thirty to forty inches in diameter and from one to three feet deep) which have been called "Indian bathtubs" at least since 1884, are debatable as to origin (Stewart, 1929). Some authorities ascribe the formation of slightly similar "weather pits" to the chemical action of water, natural acids, and ice (Matthes, 1930; Elsasser, 1962), but this does not fully explain the type of basins found at Mountain Home.

There exists one account of how Indians used these basins. It was probably written by Mrs. Belle Cramer Hassock, one of the first residents of the North Tule area, and is quoted below without vouching for the accuracy of all its details:

"The Inyo Indians used to come for acorns to Mt. View and camp across from the Cramers. The squaws shelled the nuts with their mouths until they bled. The shelled acorns were dried on the rocks in the sun, then hammered in small holes with rock pestles. Then big holes in boulders were wiped out clean with leaves until shiny, and the acorn meal mixed with water was put into them. A fire was built close at hand, and from it hot stones five to ten inches in diameter, were rolled into the holes. The mixture was kept boiling until it was thick like mush. It was then placed on rocks to dry . . . . It was broken up and then put away for winter use." (Stiner, 1934)

Rocks with basins or grinding holes in them were commonly called "mills" by the early settlers as evidenced by two newspaper accounts entitled "An Aboriginal Mill" and "Old Indian Pioneer Flour Mills of Tulare County" (with illustrations) in Stewart's Scrapbook (1933, p. 72 and p. 50), and probably by some references to "mills" found in surveyor's notes.

Harold Schutt's summary (1962) of these intriguing basins, although it does not answer all questions, offers perhaps the most plausible explanation to date of their origin. He quotes geologist G. K. Gilbert, as follows:

"A moulin, or glacial mill, is a stream of water plunging from top to base of a glacier through a well of its own maintenance. . . . At the base of the ice the plunging water finds boulders and sand, and with these, its familiar tools, attacks the rock bed. Some detail of the configuration of the bed, the presence of a large boulder held by the ice, or some other local condition, permanent or temporary, guides the water in such way as to determine scour at a particular spot, and a shallow hollow is made. As successive moulins pass the spot the hollow itself serves as a condition to determine further scour at the same spot. At the same time the hollow serves to prevent scour in its immediate vicinity, but when the moulin has moved beyond its influence another hollow may be initiated. As moulin follows moulin and summer follows summer, the hollows are deepened and assume the character of potholes. . . . After a hollow has been made and the condition for a whirl thus permanently localized, the whirl may be maintained by violent motion of the water anywhere about its rim; so that the deepening of the pothole progresses whenever a moulin stream strikes near it. If a moulin stream of pure water strikes the divide between two potholes it may furnish power for the simultaneous drilling of both holes without eroding the partition between them.

"If the surface conditions of the glacier are such that successive moulins follow closely the same track, there may be a long row of potholes, and with changing conditions there may result either parallel rows or an irregular distribution. . . ."

Restless Natives. The white man did not sweep away the San Joaquin Indian culture all at once. The destruction that culminated in the 1850's and '60's was preceded by almost a century of infiltration by wanderers, renegades, trappers, and prospectors of many nationalities and a bewildering variety of virtues and vices. By 1830 men like Pegley Smith (not Bible-toting Jedediah Smith) were teaching them to steal horses and drink whisky, a drinking-driving combination that was bad medicine long before the days of gas-driven horsepower.*

The early trappers and prospectors had some cause to fear the Indians of the Tule and Kern River mountains, especially the Paiutes. But apparently this danger was not taken very seriously by those who had business or sought diversion in the mountains. One evidence of this is a letter written in August,

---

*The Yokuts Indians stole horses for food; not primarily for riding. One Spanish punitive expedition reported finding the remains of over 300 horses that had been eaten or "jerked" in a few days at one encampment.

1855, from "Green Horn Gulch, Kern River," by Thomas M. Heston, pioneer Tulare County business man, in which he tells of a previous hunting trip on Bear Creek. Very probably this was the Mountain Home Bear Creek because he locates his camp as being forty-two miles from Visalia. He describes a fracas between the five hunters and 300 Indians, with the hunters making the Indians "vamoose the rancherie," according to his account. He tells of killing an antelope every day for fresh meat at their Bear Creek camp. Heston wrote of other Indian disturbances and says, "The people in the Valley fear an attack and if it comes, there will be hard fighting." (Tulare County Historical Society, 1958)

By 1856 the Indians and whites in Tulare County were continually "in each others hair," sometimes with fatal consequences. (In fairness to the Indians it should be said here that it was a few whites, not the Indians, who collected scalps.) The situation was ripe for some incident to set off a campaign to exterminate the Indians or put them on reservations.

<u>The Tule Indian War.</u> This last-ditch stand by the Indians on Tule River has received little attention either locally or among students of Indian history. We tend nowadays to treat it as a comic-opera affair, but it engaged most of the able-bodied settlers in Tulare County, and all available soldiers from Millerton to Fort Tejon. The mobilization was heavy but the fighting light. Several histories of Tulare County, including those of Barton (1874) and Menefee and Dodge (1913) include accounts of it, but there is no reason to believe they are more accurate than the stories of John Barker (1955) who was there, and Abraham Hilliard who wrote about it in 1860, four years after the event. The following is from Hilliard's account as reprinted by Thompson (1892):

"The Indians of Tule River, always the most hostile of any of the region, stole a band of 100 head of cattle from Frazier Valley.*. . . . They were driving them to the headwaters of the Tule River when overtaken by the owners. The Indians declined to surrender; said they were driving the cattle to their battle-ground, and meant to fight the Americans. . . . A

---

*Barker says they were from Elisha Packwood's herd of "thoroughbred" Kentucky cattle. Packwood, a well-heeled resident of San Jose, had just established a thriving ranching operation in the Porterville-Lindsay area.

THE TULE INDIAN WAR 15

company of sixty volunteers was organized from different parts of the county. David Demasters was selected as captain. . . . [The] company sallied forth from Visalia, and in the foothills encountered bands of Indians who fled at the approach of the whites. They were pursued some thirty miles to a locality since called 'Battle Mountain,' on the North Tule, where most formidable fortifications were encountered, exhibiting remarkable skill upon the part of the savages, and their method of retreat and their fighting showed a strategy and bravery worthy of civilized admiration. This position was selected where an almost impenetrable thicket of scraggy manzanita and live-oak chapparral, extending a long distance, protected the flanks and rear, and a stone wall five feet high, and brush abattis, protected the front or sally-port. This was in a region of rocks and boulders, and caves and fissures, so thick with brush that the attacking force could not see a rod, although the smoke from the Indian camp had been visible from a distance. There were undoubtedly secret paths of ingress and egress to this defensive position known to those within, but not to those without. Assaults were attempted through the brush without success, three whites being wounded by arrows from unseen foes. Three Indians had been killed outside the fortress, but those within appeared safe. After two days' siege, and several ineffectual assaults, the company retired, and reinforcements were sent for."

The remains of this "fortress" is still discernable on the Harry Scruggs ranch, and he has a three-inch cannonball that was found on the hill across the Balch Park Road to the south. Hilliard goes on to describe a second assault by a force of 140 men under Captain W. G. Poindexter, the additional men having been recruited from Visalia and the new mining district at Keyesville on Kern River. This is described as a "desperate battle" but indecisive, and the whites returned to Visalia. Some later accounts indicate that the fighting prowess of the white volunteers was held up to considerable ridicule in the saloons of Visalia. Hilliard continues:

"The fact and seriousness of the Indian War had been widely known, and reinforcements came to defend the settlers of Tulare and to share in the danger and glory. There were volunteers from Mariposa and Merced under command of Captain Ira Stroud and Captain John L. Hart. The most important assistance, however, was a detachment of 25 United States soldiers from Fort Miller under command of Captain Livingston, having a small mountain howitzer, and half a company of United States cavalry from Fort Tejon under command of Alonzo Ridley, Indian Agent. Altogether there was now an army of 400 men under command of Captain Livingston. The Indians were reported to be 700 strong, but of course there were women and children in the number. This force marched to the attack of the Indian stronghold. Captain Livingston did not order any assault, but taking his regular soldiers with his howitzer ascended a small hill where he could drop a few shells among the Indians in the fort. This was a new method of warfare to the untutored but brave and skillful red warrior, and he indiscreetly made a sortie to capture the small party with the big gun. As they came forth they were met by Captain Livingston and his men who repulsed the attack and charging upon the enemy entered the fort and the Indians fled as best they could. Some of the officers were wounded in the first attack by the Indians, but in the charge no white man was hurt. It was estimated that 100 Indians were killed. The Americans pursued the Indians for two or three days but they scattered and concealed themselves in the mountains and the pursuit was abandoned. The volunteers returned to their homes. The regular troops guarded the foothill passes for a few months, but the Indians succeeded in eluding them on several occasions and committed some depredations.

"An appeal was then made to the Indian department to take care of its wards, and Sub-Indian Agent William Campbell from Kings River Agency, was induced to seek the Tule River Indians and make peace with them. He visited them in the mountains where runners had been sent to collect leading men of the tribes, and a treaty was made restoring peace, which has since continued. This war and this thorough chastisement taught the Indians a lesson of the strength and determination of the white man's government, which they very much needed, and which resulted in great benefit to all. The country then prospered more rapidly; explorations were safely conducted in the mountains, the mountain valleys were settled, sawmills built and many developments made.

"The Legislature of 1857, asked of Congress an appropriation of $410,000, to pay the Indian War claims of California, of which $10,000 was for the Tulare War of 1856."

THE TULE INDIAN WAR

An interesting character of this period was Orson Kirk "O. K." Smith who, according to Barker, was the leader of one of the first expeditions against the Indians in June 1856. A claim in his behalf was submitted against the government for $15,000 for a sawmill and 100,000 board feet of lumber burned by the Indians (Mitchell, 1962). In apparent contradiction to this, Barton's (1874) history represents Smith as a friend of the Indians and says his mill was protected by them while they burned the property of white men who had been unfriendly to them. This must have been the first sawmill in what is now Tulare County. Its location has not been definitely determined.

Barker (1955), as an eyewitness, has some interesting sidelights on the final assault of the "war":

"We left Visalia and entered the mountains through and up the Noqual [Yokohl] Valley and with the aid of saddle horses, men and ropes, we 'man-handled' that gun up over some very steep mountains, and finally set our camp about a half a mile from the Indian Fort. The next morning we made a reconnaissance in force in order to draw out the enemy, and to force a plan of attack. They climbed their breast works, and reviled and defied us in the vilest and filthiest manner in Spanish expletive. We retired without making an attack, and our officers held a council of war to decide on how the final assault was to be made.

"In the morning as soon as we had breakfast, and after throwing a few shells into the fort, we marched up in front and between the two horns of the crescent so that they had a cross fire on us from the horns on each side.

"Several of our men were struck with the arrows, they had no guns, and the arrow was very effective at such short range.

"The lieutenant climbed up on an immense boulder as large as an ordinary house so as to look over their wall. Although it was in the month of June it was quite cold at night and early in the morning.

"The lieutenant had a military cloak over his uniform and they made a target of him. We saw the arrows strike him several times but they could not penetrate the cloak and being shot from an angle below they simply stuck in the cloak and slipped up and hung there. It seemed as though one must have stung him for he commenced to swear, and ordered his men to charge the breast work. Upon this we all went in and in about ten minutes we had it all our own way. There were forty

Indians dead and how many wounded we could only surmise. The squaws made their way up through the canyon, following the bed of the river. We immediately commenced to loot the stores. There was a great quantity of dried beef made from Packwood's fine cattle, stores of pine nuts, arrows, grass-seed, and grasshopper cheese.* There was the plunder they had stolen from the houses they burned, saddles, etc., and such a store of Indian baskets as would today delight the heart of a connoisseur, all of which were condemned to the flames.

"Thus ended the Indian war of 1856.

"We followed them through the mountains for nearly two months after this but no more were slain. . . ."

The Tule Indian War, however, did not end the Indian disturbances. They merely moved east into the Owens Lake country. W. A. Chalfant (1922) says, "The Indian population of Owens Valley was augmented in 1859 [the date probably should be 1856] by fugitive Indians from Tule River. . . . [which] sent into this county numbers of Indians with a ready-made and burning hatred of the white man." The Owens Valley war broke out in the winter of 1861-62 and there were many casualties on both sides. One of the final episodes took place in April 1863 at Whiskey Flat (now Kernville) where thirty-five non-combatant Kern River Indians were killed by soldiers.

Summary. The first men of Mammoth Forest have no names. They had names, of course, when they were alive. Mothers and wives must have handles to use to call their children to supper--and their men to account. But to us, even the name of the war chief who led the Yaudanchi and Paiutes at Battle Mountain is already "lost in the mists of antiquity."

Since our early historians were so negligent in recording even the names of their Indian enemies, we most borrow words from another time and place with which to conclude this chapter. Joseph, the Indian war chief of the Idaho Nez Perce, in a situation very similar to that of the Yaudanchi after their defeat, spoke thus his "farewell to arms" at the tribal council:

"I am tired of fighting. Our chiefs are killed. . . . The old men are all dead. It is the young men who say yes or no. . . . It is cold and we have no blankets. The little children are

---

*Any ladies wishing the recipe for grasshopper cheese may get it from Uncle Jeff Mayfield's book (1929).

SUMMARY

freezing to death. My people, some of them, have run away to the hills and have no blankets, no food. . . . I want to have time to look for my children and see how many of them I can find. Maybe I shall find them among the dead. Hear me, my chiefs. I am tired. My heart is sick and sad. From where the sun now stands I will fight no more forever." (Fuller, G. W., A History of the Pacific Northwest)

CHAPTER IV

FIRST DISCOVERIES, TRAPPING, MINING, AND HUNTING

> So long as I have legs, so long as I have eyes, wherever I go I am lord of the mountains and rivers and the winds and the breeze.
> *One Hundred Proverbs,* translated from the Chinese by Lin Yutang.

AT LEAST A CENTURY has now passed since the first white man stood and looked up in silent wonder at the redwoods of the Southern Sierra. Who this man was we do not know. But many are those who have followed him—and marveled at the unbelievable bulk and beauty of the great red trees.

Even the relatively remote North Tule country was probably visited by some white wanderers before the Gold Rush Days. They were not necessarily Spaniards because these people did not penetrate very far into the Sierra. But the American fur-trappers ranged far and wide through the Sierra following Jedediah Smith's visit in 1827, as has been previously described. It is quite possible that some of these restless mountain men checked the meadows and streams of the Tule River redwood country.

Frank Knowles. Among this hardy tribe of fur-trappers was one man who came early enough and stayed long enough on the upper Tule to gain the reputation of being the "discoverer" of that country. This was Frank Knowles, called by his friends "the old hermit of Bear Creek." It has been established by the Tulare County Historical Society (1955) that Knowles was trapping in the Camp Wishon area at least as early as the floods of 1861-62.

Some interesting details of Knowles' later life are recorded by the Tulare County Historical Society, including his service as a member of one of the three parties that first climbed Mt. Whitney in August and September of 1873. He is described by King (1915) as being "a sort of chamois" because of his mountain climbing ability. Local people have described him as "a short man with a long rifle" and a flowing beard.

Although Knowles lived alone he was not unsociable. Everyone in the area knew him and he was apparently everybody's friend. He is reputed to have been one of the fiddlers for the regular weekly dances at Doty's Mountain Home resort and to have made good violins from manzanita and redwood. Besides apples, he raised strawberries and pansies at his clearing in Section 2 near the present Bear Creek Road. Mrs. Ola Doty Hubbs of Visalia tells how he used to come to the Coburn lumber camp on Bear Creek on Sundays, announcing himself from the hill by calling, "Hello-o-o Justin," to his friend Justin Burgess, the logging boss, and then waiting for the welcoming, "Hello-o-o Frank," before coming down with his pack of dogs for Sunday dinner.

Knowles trapped "silver fox, red fox, pine marten, bear and other animals" (Tulare County Historical Society, 1955), and probably wolverine and fisher. It is believed that there were beaver in the area before Frank Knowles' time, and if so, they were the first of the wild inhabitants to be exterminated.

After the area was surveyed in 1878, Knowles took up the 160 acres around and below his cabin. All except the 10 acres containing the cabin and orchard later became part of the Sequoia National Forest. The Knowles tombstone in Crabtree Cemetery reads, "Frances F. Knowles, native of Maine, 1824-1903" (Native Daughters of the Golden West, 1954).

First Written Record. Men of another breed swarmed through the Sierra following the fur trappers. These were the prospectors. A Tulare County Chamber of Commerce publication (1959) says that miners began working south from Mariposa in 1852, and that in 1857 the "Kern River Gold Rush brought an avalanche of travel toward the mountains." Here again, the writer believes that the gravel bars and rock outcrops of the Tule River, and especially the mineralized areas along the limestone belt across the Wishon Fork, must have been sampled in the 1850's by some of these hardy and gold-hungry men. The Tulare County Historical Society (1949) reports that Dr. Samuel S. G. George of Porterville discovered the copper and galena deposits above Camp Wishon.

A mining man was the first person known to have seen the redwood forest of this area and left a written record of what he saw. His name was B. W. Farley. According to the San Francisco Daily Alta California of September 14, 1860, Mr. Farley travelled from Visalia to the newly discovered Coso silver diggings in "Eastern Tulare County" (now Inyo County) in the summer of 1860. B. W. Farley was probably a relative of the Minard H. Farley who had led a discovery party into the Coso area in March 1860 by way of Walker Pass (Chalfant, 1922) and reported his trip in the Alta of July 24, 1860. The September 14 edition quotes B. W. Farley, under the heading "Discovery of Mammoth Trees," as follows:

"After reaching the summit of the first ridge we bore gradually to the right, and travelling in an easterly direction we entered a forest of redwood timber, that astonished the natives themselves. . . . I lit down and measured one. . . . a little over 33 feet in diameter and supposed to be at least 300 feet high. . . . They point index-like to Heaven, read chapters to passers-by on the immortality of man."

Apparently the redwoods were beginning to evoke poetic thoughts even in 1860.

The measurements Farley gives of the big redwood tree are approximately those of the tree that then stood proudly on what we call the "Centennial Stump." Farley most likely traveled the Dennison Trail (see Chapter V). It is named the "Farley Trail" on "Farley's Map of the Newly Discovered Tramontane Silver Mines in Southern California and Western New Mexico," published in 1861. This rugged route traversed the Mountain Home redwood area from near Brownie Meadow to beyond Shake Camp. After traveling "among these monsters for several miles" Farley's party "commenced descending in an easterly direction to the North Fork of the Kern River," according to the newspaper account.

Old Spanish Mines. The North Tule area has no monopoly on "lost mine" legends and other old prospector's tales, but its stories about some very early pre-Gold-Rush mining activities are worth recording. Doctor (1959) says, "It is a historic fact that Spanish from the Coast prospected for and probably found gold in Tulare County before the American settlement." Such expeditions in the early part of the nineteenth century are documented in old mission records.

On April 23, 1885 the Visalia Weekly Delta reprinted an article from H. Wallace Atwell's "Trade Magazine and

OLD SPANISH MINES

Immigrant Guide" which stated that Thomas Osborn (a well-known early resident of Yokohl Valley) had located a gold mine a few years previous "immediately under the Big Tree Belt of timber on the Middle Fork of the Tule River." This was almost certainly on the stream now called the Wishon Fork or North Fork of the Middle Fork, because maps prior to 1905 show that stream as the "Middle Fork." (The one now mapped as the South Fork of the Middle Fork was referred to as "Nelson's Fork.")

Osborn, it was claimed, found the mine with the assistance of a Mexican named Don Carlos who had been told by a Los Angeles Mission halfbreed Indian that when he (the Indian) was a lad he was one of a large number of Indians taken by Spanish padres through Tehachapi Pass and north along the hills. This was some time in 1824 or '25. After several days travel, "crossing several small streams and one large one," the party made camp at the head of a stream and the Indians were set to work digging gold ore, carrying it in sacks to camp, and grinding it. After a fight with local Indians the mine was filled in and the party returned to Los Angeles.

Osborn and Don Carlos thought they found the padres' campsite and their "mill," a huge, flat granite rock twenty feet high on a steep hillside. The rock had "many deep barrel-shaped holes, worn deep from continued stamping of quartz; some six feet deep with circumferences of one to three feet." (This sounds like some of the so-called "Indian bathtubs.") They did not find the old shaft, but many years later a man named Wimer (or Wymer) claimed that he found it on top of the Moses Mountain Ridge in Section 1. "Dude" Sutch of Springville tells how Mr. Wimer about 1928 dug out an old shaft ninety feet deep and found pieces of Spanish pack saddles at the bottom.

Another interesting story of "Spanish Gold" is the following from the <u>Visalia Weekly Delta</u> of July 4, 1879 headlined, noncommittally, "Old Mine Discovered, Probably": "A few days ago Messrs. Joseph Street, James Ramsey and John and Tod Gill discovered an old tunnel on Middle Tule that had evidently been worked many years ago." A great landslide near the mouth of the tunnel had changed the river's course. The discoverers "let themselves down to the tunnel from the top of the bank by rope and explored it a distance of 63 feet." They reported that the excavation was expertly timbered.

Stories are told (by R. C. "Pop" Lloyd of Hanford, for one) that there used to be several "gentlemen of leisure" around the North Tule area who seemed to be able to go into the mountains whenever they needed money and bring back enough gold for their bacon and beans. The names of Dennison, Derasties "Rat" Wiesner, and Jim Street were mentioned.

For what it is worth, it is the writer's opinion that the more likely locations for the old Spanish mines are in the old Keyesville-Glennville-Tailholt area north of Kern River, and that they were found and worked out in the Kern River gold rush in 1854 and years following. Jose Jesus Lopez, former major-domo of the Tejon Ranch, once said, "Knowing the padres, I don't believe they ever lost anything."

Some searchers for mineral wealth did not come back. The story behind "Rose's grave" on Burro Creek above Camp Wishon was told to the writer by Mrs. Ethel Rose Hughey of Porterville, granddaughter of William Rose, the man buried there. He had taken two of his young sons, one of them only fourteen years old, and one of the Crabtree boys to work a prospect he had found, leaving his wife at "Mountain Home" (probably at Kincaid's mill). Upon arriving at the site in June 1879 he died suddenly. The boys put his body on a pack horse and traveled three days toward the Mountain Home area. One rode ahead to get Mrs. Rose. He was buried where they met on the trail. The prospect was lost. Their mother would not let the boys go back to look for it until they were of age and by that time their landmarks were grown over or washed away. One more lost "mine."

Exploration. From Spanish days onward many official exploring parties looked up from the level floor of the San Joaquin Valley and saw the timbered ridges of the Mountain Home country, but apparently none took the time to visit the area until 1867. An official government report of 1900 by Pinchot, "The Big Trees of California," devoted only eleven lines to the North Tule groves, including the statement, "We are not aware that these two groves were known previous to their discovery by Mr. d'Heureuse, one of the topographers of the Geological Survey, in 1867; at least no notice of them has ever appeared in print." (Evidently Pinchot had not read B. W. Farley's account in the San Francisco Alta of September 14, 1860, already mentioned.) Mr. R. d'Heureuse made a map for the California

HUNTING AND TRAPPING 25

Geological Survey that probably included the North Tule redwoods, but it is lost or kept in a very safe place.

The most tireless explorer of the Sierra redwoods, and the best-known writer about them, was John Muir. In the autumn of 1875 he set out to walk the Sierra from north to south to find the southern limits of his beloved "giants." After visiting Hale Tharp in his "noble den" in the log in the Giant Forest country and spending several days studying and admiring a "grand fire" in the redwoods in the Kaweah drainage, he entered the North Tule area.

Muir was not one to clutter his writing with geographic details about his travel routes, but he did mention the "terribly precipitious canyons encountered in crossing the Tule basin," and more than once repeated his conclusion quoted at the beginning of Chapter I, that the North Tule redwoods were the finest in the Sierra.

Muir's known writings (1878, 1901) are strangely brief from the time he left the Kaweah forest fire to his finding of a sawmill on the South Tule. (The mill was undoubtedly Porter Putnam's, which was running full blast that year according to the June 17, 1875, Visalia Weekly Delta.)

On October 20, 1875, Muir (1938) wrote in his journal, "Here is temple music, the very heart-gladness of the earth going on forever. On the Middle Fork of Tule I found a Sequoia forest 8 miles long, six wide, and wedge shaped. . . . I saw flocks of ladybirds going into winter quarters."

Hunting and Trapping. In this period before 1884, hunting was sometimes done for sport (see Chapter III) but it was also a necessity for food and for protection of cattle and sheep from predatory animals. Trapping was a way to obtain hard cash, often a scarce item. Some contemporary quotations indicate the attitude toward hunting and hunters:

"Splendid deer skins, dressed, were offered for sale . . . . at $19.00 a dozen." (Delta, October 20, 1861)

"Two hunters living in the foothills on the waters of the Tule River, have killed over 120 deer during the present winter." (Delta, 1861)

Stiner (1934) says that market hunters killed four or five dozen quail per man per day for several years north of Yokohl.

Alvin H. Slocum, North Tule pioneer, "tells of the killing of 14 or 15 bears in the autumn of one year and relates how in one hunt he shot 21 bucks; his largest bear he killed in 1867. Jacob Cramer, Marvin Wilcox, and Frank Knowles were with him and they have often testified that it weighed, dressed and without hide or head, 1550 pounds." (Menefee and Dodge, 1913) Probably any one of these expert witnesses, after verifying Slocum's story, could have told a bigger one. King (1902) says that the hunters that he visited in the Giant Forest country in 1864 "went on in their old eternal way of making bear stories out of whole cloth."

The last antelope in southeastern Tulare County is recorded by Stewart (1956, p. 55) as having been seen in 1875 near Springville.

Even after 1915 Jay C. Bruce, the famous state-employed lion hunter killed 700 mountain lions in California, according to The California State Employee of March 1963. Irvy Elster of Springville assisted him in killing many of them in the North Tule forests.

CHAPTER V

TRAILS ACROSS THE SIERRA

> We sing a song of mountain trails
>   Through sunny beds of heather:
> We chant a hymn of forests dim
>   Where the tall trees sing together.
>
> We sing the ageless, endless song
>   Of fur and fin and feather,
> Of rivers clean, and meadows green
>   Where the deer trails come together.
>
> We jointly hail life's sacred source,
>   Praise sun and stormy weather;
> And celebrate our common fate,
>   The hills and I together.
>
> Floyd L. Otter, "The Hills and I."

ANY HOUND that sniffs footprints made a hundred years ago is on a very cold trail; but for the old Sierra trails a little scent remains.

It is a commonplace observation that highways tend to follow the routes of earlier roads, that these were once horse and livestock trails, and before that they were Indian paths. Indians, of course, followed the trails made by game. The Indians left us no more maps than the wild animals, but there are some hints that the first white men to cross the Sierra in Tulare County were guided by Indians over their ancient routes.

The main concentration of Indian population in the southern San Joaquin was on the Kaweah delta and around Tulare Lake; and the heaviest Indian population on the opposite side of the Sierra was around Owens Lake. These two groups did a good deal of trading (Latta, 1949), and used trails for hunting, food-gathering, visiting, and just "going camping." The evidence indicates strongly that main-travelled trails from Kaweah and

Tule villages led to a common meeting ground near the junction of the Kern and Little Kern Rivers (Round Meadow-Trout Meadows area). If you wished to continue eastward you would travel one of several trails that converged on this meeting place from several Owens Valley points.

The white man's trail work tended to follow this same pattern, with all trails converging at Trout Meadows. P. M. Norboe (1903), a well known surveyor, reported, "There are four well beaten trails entering the valley of the little Kern from Tulare Valley and all unite before reaching the Big Kern." The best route was described as the one through Camp Nelson; the roughest, up the South Fork of the Kaweah. A Porterville party taking a trip to Mt. Whitney in 1878 was described by Anna Mills Johnston in the first issue of the Mt. Whitney Club Journal. She said their route was by way of Dillon's Mill over Chisel Mountain, to the Little Kern without a trail, down that stream to Martin Click's sheep camp, and thence to Trout Meadows.

The Indian "foot paths" (see the account of Hale Tharp's discovery of Giant Forest by Doctor, 1959) were soon modified to suit the needs of the packers, who with their pack mules and horses carried supplies to sheep camps, mines, and other mountain activities. Still later many of the routes were converted to roads and even to paved highways. For instance, the trail (sometimes still referred to as the Old Indian Trail) that connected the Yaudanchi village northwest of Battle Mountain with Balwisha villages on the South Fork of the Kaweah has become the approximate route of the present road over Blue Ridge through Grouse Valley. Another trail undoubtedly connected the main village of the Yokod (Yokohl) tribe, near the present city of Exeter, with the Yaudanchi villages at Rancheria and Milo. This was very likely the route chosen by John Jordan for the western section of his trans-Sierra trail.

The Bear Creek and Balch Park roads apparently follow quite closely the Yaudanchi-Paiute food-gathering and trading trails, as evidenced by numerous Indian grinding holes where these routes cross the black oak belt. Yokuts traders took deer and antelope skins, soapstone, saltgrass salt, and baskets to Owens Lake and traded these articles to the Paiutes for common salt and obsidian for arrow points (Latta, 1949).

THE DENNISON TRAIL

<u>The Dennison, Coso, or Farley Trail.</u> Stop some spring day along Highway 65 between Strathmore and Lindsay in Tulare County and look northeastward toward the High Sierra. Their western outposts will still be gleaming white with the winter's snow. Now, mentally "time-machine" yourself back to a corresponding day in 1860.

Word has just reached Visalia that rich silver strikes have been made just east of that mountain range, southeast of Owens Lake in the Coso Mountains. Let your eyes follow the Sierra skyline from north to south. Slightly north of due east from where you stand there appears to be a gap in the "garden wall." This is the timbered ridge now known as the Mountain Home area, lying just south of Moses Mountain. A little further south the mountains rise again where Jordan Peak presents its snowy western face.

Men of the Visalia-Porterville area who heard about the strike probably looked at the Sierra from the Stockton-Los Angeles Road (later the route of the Butterfield Overland Mail and now part of Highway 65) and wondered if there wasn't a shorter, cooler, better-watered route to the Coso silver country than the previously-used Walker Pass route. W. A. Chalfant (1922) says that, by July 1860, parties were leaving Visalia almost daily for the Coso diggings. Many of these parties went by this short-cut route.

B. W. Farley's description of this route indicates that he followed a trail elsewhere called the Dennison (or Denison, or Coso) Trail. This trail apparently preceded the now better-known Jordan and Hockett trails across the same general area. Very little is known about it. Still less is remembered about the man for whom it was presumably named. Mr. Tillman Phariss of Porterville says Dennison was a mountaineer who was killed near the old North Fork waterpower sawmill by accidentally tripping the wire on a set-gun he had rigged up for a bear. He was buried on the spot. There is a Dennison Mountain, Ridge, Peak, and Ditch, and one of the first school districts in Tulare County was the Dennison School District.

According to Farley it was about 95 miles from Visalia to the Coso mines by this route. Menefee and Dodge (1913) quote the <u>Visalia Delta</u> in 1861 that, "Captain George, an Indian, and big Injun heap at that, has commenced running as an expressman between this place and Coso. He makes the trip now in about four days." A Captain George, the following year, was one of the leaders of the Indian uprising in the Owens Lake country. Perhaps he combined mail-carrying with war-mongering among the mountain tribes.

The route of the Dennison Trail can be deduced from references to it in a toll road application made in 1870 (see Chapter VII), from the official Tulare County map of 1884, from Farley's map mentioned in Chapter IV, and from the memories of a very few oldtimers. Mr. Irvy Elster of Springville has a remarkable memory for stories handed down from his father, many of which can be checked only by reference to old records. He says this trail left the old Butterfield Overland Mail Route between the locations of Strathmore and Farmersville, went through Spanish Camp (which is on the divide between Yokohl and Lewis Creeks), thence south and east past the old Springville rodeo grounds on the Clemmie Gill ranch, crossed the North Fork of the Tule River near the mouth of Sycamore Creek, and then passed northeast over McDonald Hill to Rancheria. From Rancheria it climbed the main ridge between Bear and Rancheria Creeks through the Otis Lawson ranch to the Balch Park Road near Brownie Meadow (Farley's "summit of the first ridge"). From there it followed approximately the route of the road to Shake Camp; then down the ridge to the east, crossed the Wishon Fork, climbed a ridge south of Silver Creek, crossed the Tule-Kern divide over Maggie Mountain, followed the Little Kern down to the Trout Meadows area, and then eastward to Olancha and Coso. This is the route described to Elster by his father who took sheep over the trail in 1879. Mr. Otis Lawson of Springville and Mr. Malvin Duncan of Porterville, who operated pack outfits in that area before 1920, corroborate this routing as far as Shake Camp. The "Dennison Trail" past Summit Lake as shown on the official county map of 1884 may have been a change made later to improve the trail.

As a side-light on the trails through this area, there is a considerable body of tradition about a traffic in stolen horses. This traffic probably reached its peak around 1870. It is said that horses were stolen in the San Joaquin Valley and driven by approximately this route to the Owens Lake country; and there, for the sake of efficiency, stolen Owens Lake horses were brought on the return trip to sell to the San Joaquin people who had so unfortunately lost their means of transportation. The names of Tiburcio Vasquez, Frank Ryan, and Ryan's

THE JORDAN TRAIL

son-in-law, Ed Diaz of Porterville, have been mentioned in connection with transportation of horses, not necessarily stolen ones, via this route.

The Jordan Trail. The story of John Jordan, his dream of a road across the Sierra from Visalia to Owens Valley, and his tragic death at Kern Flats make a fascinating story. More legends and misinformation are available about the Jordan Trail than any of the other routes through this area.

B. W. Farley, in his 1860 letter already quoted, wrote, "a good pack trail can be made, over which ordinary mules can pack 250 pounds, at an expense not to exceed $800-1000" from Visalia to the Coso Mines. There was a lot of verbal and political activity about this time over the matter of building roads across the Sierra. Doctor (1959) wrote that,

"In April, 1860 the County Supervisors appointed a committee to view a road and pack trail from Visalia to Mono" and the Visalia Delta announced on January 30, 1860, that a company had been formed and application made to the legislature to build a toll road from Visalia to Owens Lake. Included among the men interested in this road were Samuel S. G. George, H. L. Matthews, S. Sweet, Henry Bostwick, and John and W. F. Jordan. Chalfant (1922) says the company was granted a charter by the Legislature of 1862, the road to start "between Deer Creek and Kings River. . . , thence across the Sierra Nevada mountains to a point between the north end of Owens Lake and the north end of Little Lake."

While all this talk and legislative action was going on, John Jordan and his son, William F. Jordan, were apparently going ahead with a trail project under authority given by the Tulare Board of Supervisors. The right-of-way was thirty-three feet wide and the trail was to be completed in two years. The Jordans did not petition for a wagon road right-of-way but the Board of Supervisors wrote into their agreement that a wagon road was to be completed within five years. The Jordan petition is on file at the Tulare County Courthouse. The full text is as follows:

"Your petitioners would respectfully petition your Honorable Board to have made and declared open, a pack trail or passway leading from Tulare Valley across the mountains to the South end of Big Owens Lake, said trail to commence at George E. Long's residence (near the site of Exeter), passing through Yokall Valley and thence Easterly across the mountains to the said lake, and would further petition your Honorable Board, that the right-of-way and privilidge of constructing the said trail be given and granted to John and William F. Jordan, the applicants and petitioners herein, and for whom your petitioners will ever pray —."

It included twenty-two signatures (including that of J. B. Hockett) and was filed November 1, 1861. Early in May 1862, Jordan told the Editor of the Visalia Weekly Delta that his trail would be open to pack stock by June and to wagons a short time thereafter. But only a month later he lost his life in the snow-fed waters of Kern River (Doctor, 1959) while he and his sons, Allen and Tolbert, and a man named Gashweiler were attempting to cross Kern River on a raft at Kern Flats.

Jordan's sons did not carry on the project, but according to Menefee and Dodge (1913) $1600 was raised by subscription in Visalia, and G. W. Warner finished the trail including the building of a bridge across Kern River some 50 miles from the nearest wagon road.

The Delta says the trip from Independence to Visalia via this trail took three days. The soldiers of the Union Army used it in 1863 in their move from Independence after the Indian trouble there and on upper Kern River (Doctor, 1959; Chalfant, 1922). In 1875 the Wheeler expedition ("Exploration and Surveys West of the 100th Meridian") traveled eastward on the Jordan Trail, according to a message discovered between Mountaineer and Clicks creeks, and reported in the Fresno Bee of September 28, 1926.

The various routes of Jordan's trails are not easy to pin down. Norboe (1903) states, "The Jordan Trail was originally constructed to run from Tule River above where Globe is now, past Rancheria and Bear Creek, through the forests at the foot of 'old Moses' where the charming summer resorts of Summer Home and Mountain Home and the Enterprise Mill are now located. Then it crossed the Middle Tule near Doyle's Soda Springs, and by a very steep, rough, and by no means a safe route, climbed the mountain to Jacobson Meadow." From this meadow it was a relatively good trail to Trout Meadows, and thence, according to Menefee and Dodge (1913), it went "up Big Kern to a point below where Kern Lakes now are, crossed the river and proceeding eastward via Monache Meadows, was to strike Owens River below the lake."

THE HOCKETT TRAIL 33

The earliest map found that shows any of the Jordan Trail routes is the government survey township plat of 1878 by P. Y. Baker. It differs from Norboe's routing in that it shows the Jordan Trail following the approximate route of the present Bear Creek road past the section line between Section 10 and 11, and thence easterly toward the river crossing last mentioned above, thus missing Balch Park by over a mile. Claud Jordan of Visalia says his grandfather had a summer route through the higher country and a lower route to use when the snow was deep.

The western terminus and headquarters of the Jordan Trail was his ranch near Rocky Hill east of the site of Exeter. He brought his freight by "bull-team" from Stockton, according to Claud Jordan, and transferred it to pack animals at this ranch. John Jordan came originally from Pennsylvania and had seen service in Texas with Sam Houston in the Mexican War before coming to California. He had experience as a surveyor and is said to have been more interested in an eventual wagon road across the Sierra than in a toll train. He had twelve children and his descendants are numerous in California. (He is not to be confused with his nephew, Capt. John F. Jordan of slightly later Tulare County history.) A redwood tree in Balch Park has been officially named for John Jordan.

The completion in 1864 of the toll road from the San Joaquin Valley to Owens Valley by way of Walker Pass apparently put an effective damper on the trans-Sierra trail-building activity across the Tule River country. This road was built by the McFarlane Toll Road Company.

The Hockett Trail. The Tulare County Board of Supervisors on December 11, 1862, granted to Henry Cowden, Lyman Martin, and John B. Hockett permission to build a pack trail commencing "at a point in the Tulare Valley near where the Kawiah River leaves the foothills," and thence easterly across the Sierras to the "foot of Big Owens Lake between Haiwee Meadows and the Lone Pine Tree" (now Line Pine, Calif.).

On August 5, 1864, Cowden presented a sworn statement that the three men above named had completed the trail at a cost of $1000 and asked permission to charge tolls. The supervisors thereupon set toll rates of 50 cents for a mule or horse, 25 cents per head of cattle, 5 cents per sheep or hog, and 25 cents for a man on foot. Doctor (1959) says Hockett built this trail while working for the Army.

This trail became the best known of the old trails. Dyer (1898) said it was marked by "peculiar blazes," and that, "the Hockett Trail was made in early days and today it remains a plain, well-blazed track from Lone Pine through to Visalia."

An alternate route led up Yokohl Valley and the North Fork of the Tule by way of Dillon's Mill. This route, according to the Visalia Weekly Delta of July 2, 1874, was "the most direct route to the new mines in the Mineral King and Little Kern districts." Mr. Ira B. Dillon advertised to take passengers by this route for $3.00 from Visalia to Mineral King (two days one way), and freight for $1.00 per hundredweight from the Dillon toll gate to Mineral King. Among the freight packed in was a sawmill. A Mr. Wilcoxon was reported to be undertaking to put up a house and corrals "at what is known as Jackass Flat just below Dillon's Mill for the accommodation of travellers." Both Elster and Mr. Malvin Duncan of Porterville say this trail went north over Chisel Peak to Tuohy Meadows. N. P. Dillon must have had a hand in it because he petitioned the Board of Supervisors in 1877 for toll road rights stating that he had already completed five and a half miles of wagon road and three and a half miles of "good pack trail."

In January 1875 the U.S. Government advertised in the Delta for bids on carrying the mail from Porterville to Mineral King via Pleasant Valley and Soda Springs (now Springville).

Clarence King on his way to make his first attempt to climb Mt. Whitney from the south in 1864 followed the Hockett Trail. Menefee and Dodge (1913) wrote that the Hockett Trail commenced "at Three Rivers, proceeded up the South Fork of the Kaweah, passing the Hockett Lakes and Meadows, and joined the Jordan Trail, continuing on its route to Big Kern (River). Instead of crossing the river at the same point, however, it continued up the stream to a point near the lower Funston Meadows, whence crossing and ascending the wall of Kern Canyon it made its way via the Whitney Meadows to the crossing on Cottonwood Creek near the lakes and thence down to Independence."

Other Trails. There are several other named trails in our area. One which was in use when the area was surveyed in 1883 ran from the Wilson (later Kincaid) sawmill site easterly,

OTHER TRAILS 35

passing northwest of Shake Camp, higher up on Moses Mountain than the present road. It can still be found at a few locations described in Norway's survey notes where it crossed the section lines northwest and northeast of Camp Lena. This trail has been referred to as the "Tuohy Sheep Trail" (Brown, 1923). It could well have been built or used by John Tuohy for packing salt to his sheep in the Tuohy and South Fork Meadows country. It was apparently an extension of the Kincaid Mill and Centennial Tree roads.

Sheepmen who went into the mountains between about 1860 and 1900 used these trails and marked out additional ones. Later the U.S. Forest Service built a very extensive trail system. The trail from Shake Camp to Alpine Meadows was rebuilt by Art Griswold about 1930 in connection with his pack station. He reports that he found evidence of an older trail, most likely the Dennison Trail.

The first pack trail up the river into the Camp Wishon area was probably the one built by J. J. Doyle in 1891 (Tulare County Historical Society, 1950) to give him better access to his homestead there than his former trail via Bear Creek and Summer Home. The Porterville Enterprise of November 29, 1907 recalls, however, that "In the winter of 1863 Frank Knowles and a trapping companion were at Doyles Springs when a heavy snowstorm came. They could not go over the Summit by way of Summer Home. So they cut their way down to the Forks, and thus was made the trail that is now the route of the [newly opened] Wishon Wagon road."

CHAPTER VI

SHEPHERD'S EMPIRE

The indefatigable shepherds have camped everywhere. . . .
Clarence King, *Mountaineering in the Sierra Nevada.*

DURING THE PERIOD 1860 to 1884 livestock grazing was by far the most important use made of the North Tule forest area. And sheep were the most important class of stock. In fact, the sheepmen of that era built up a pattern of land use that continued until after 1900, a period of over 40 years. Anyone who wishes to understand the attitudes and feelings of California-born ranchers toward their mountain lands should remember that this "shepherd's empire" (Towne and Wentworth, 1945) probably settled more nearly into a traditional pattern of land use than any system since that time.

The recorded figures on the number of sheep that grazed the California ranges in those years are hard to believe. The Visalia Weekly Delta for June 3, 1875, says, "Mr. S. S. Coburn informs us that some 50,000 head of sheep have been driven across the North Fork of the Tule River into the mountains." Mr. Coburn, father of Avon Coburn, the Springville lumberman, lived near the river crossings of the roads to both the Dillon and Kincaid sawmills, and the sheep he mentioned were undoubtedly headed up through the Dillonwood and Mountain Home areas.

The sheep population of California increased from 262,000 in 1850 and 1,000,000 in 1860 to 6,000,000 "within a few years" (Bancroft, 1884-90). Between 1850 and 1860, 550,000 sheep were trailed by all routes to the Mother Lode diggings. Sheep were a standard item of provisions for exploring and work parties as they furnished their own transportation and needed no refrigeration.

Mountain Home land probably attracted a few cattle, sheep, and horses before 1860, but the "Great Drouth" of 1862-65 brought great numbers of all classes of stock into the mountains

SHEEP

(Cleland, 1951). King (1902) said that about 4000 cattle roamed over the Giant Forest area in what is now Sequoia National Park when he was there in 1864. "Fleeing before the continued drought of the plains, all the cattlemen of California drove the remains of their starved herds either to the Coast or to the high Sierra."

The last surge of grazing in the Sierras was during the drouth of 1897-99. Dudley (1899) says that in 1898, "It was estimated that 200,000 sheep had swarmed through the Tule region alone. . . . Sheep, horses, milch cows and even pigs were frequently seen in the forests and on the meadows above 10,000 feet."

Sheep. Almost all of the information available about early grazing in the North Tule forest area is about sheep. Huffman White is credited with being the first to take sheep into the mountains of this area. He set up in the sheep business in Frazier Valley in 1859 assisted by his young stepsons, Clinton T. and William W. Brown (Stiner, 1956).

This was a favorable time to start in the sheep business because the Civil War, which started in 1860, cut off the supply of cotton and caused a sustained boom in wool and sheep prices, corresponding to the great cattle boom following the 1849 gold rush. Sheepmen reaped fantastic profits.

In the drouth year of 1864 White and the Brown brothers took their sheep to Hossack Meadow, using the Jordan Trail from the vicinity of Rancheria. At other times they used the old Dennison Trail through Brownie Meadow. They "sheeped" the area from Camp Nelson north to Sheep Mountain (near Summit Lake). They ran from 2000 to 6000 sheep, according to Clinton's son, Mr. Jay Brown of Porterville, but never more than 2000 to the band. Two herders with four dogs looked after each band. Clinton supplied both herders and sheep with salt and they pretty much "lived off the country" otherwise. The herders used a burro to move their camps.

Jay Brown says that his father told him that, to improve the range, they kept fires burning all summer in rotten logs and such places after "sheeping" around them to keep the fires from spreading. Sheepmen generally deny that they set fires to the forest or brush as they came out of the mountains in the fall, branding this story as a libel spread by cattlemen to discredit the sheep industry among conservationists.

The drouth of 1877, following soon after the 1872 collapse of wool prices (Cleland, 1951), brought disaster to sheepmen, but by 1880 Clinton Brown was back in the business and stayed in it until at least 1903. After 1881 Brown ran cattle as well as sheep in the mountains.

Another rancher who grazed sheep in the area was John Tuohy, prominent Tulare County pioneer and conservationist, whose name is perpetuated in Tuohy Meadows on the South Fork of the Kaweah. He took 6000 head of sheep into that area in 1870 and annually thereafter for many years (Tulare County Historical Society, 1957).

Other sheepmen who used the North Tule forest country included John and Newton Crabtree who sold out in 1871, J. H. Cramer (the pioneer of the North Fork Valley), Cramer's son-in-law John Hossack, and Hossack's Scottish stepfather Andrew Ross (Stiner, 1956). Mrs. Ola Hubbs says that sheepmen who packed salt out of Mountain Home after 1885 included Harry Quinn, Tuohy, Clinton Brown, Henry Zimmerman (who operated on a rather large scale around Hossack Meadow in 1896-97 according to Jay Brown), and several French (or Basque) sheepmen.

Hogs. The staple diet of the Indians was acorns. When the white settlers came they built their livelihood to a great extent on this same resource, but by means of the hog. Half-wild boars, sows, and pigs harvested the acorn crop and supplied pork to the community from earliest times to as late as the 1920's.

In drouth years such as 1897-99 valley ranchers drove their hogs to the high mountains. Clarence King on his trip from Visalia to the Mt. Whitney country in July 1864 wrote in a letter to J. D. Whitney: "I rode until nine in the evening, when we came to the 'Hog ranch,' two acres of tranquil pork," near a meadow in "the most magnificent forest in the Sierras" (Wilkins, 1958). This "ranch" (most likely only a temporary drover's camp) was on the South Fork of the Kaweah, possibly in Hockett Meadows. King later described this pig-herd in the words of its owner as "The pootiest hogs in Tulare County — nigh three thousand." Mrs. E. E. Eldridge of Fresno says her father, Robert T. Sharp, took his cattle and hogs into the mountains from their Visalia ranch during the drouth of 1898. They turned their hogs loose among the oaks on Blue Ridge, but lost nearly all of them. The Sharps camped on their timber claim near Mountain Home, a 160-acre property that was sold later to Coburn and more recently to A. C. Lindley and others. Griggs (1955) gives details about hog-raising on the open range.

Cattle. James R. Hubbs ran cattle in the mountains from his ranch headquarters located on Sycamore Creek (Greer Ranch), and later where the U.S. Forest Service Ranger Station was built in Springville. There were many others. For instance, the first of the Gill family, Levi Gill and wife, settled in Yokohl Valley in 1872, and with the help of some of their sixteen children built a cattle "empire" in the Tule and adjacent drainage areas.

George W. Wray is credited (Menefee and Dodge, 1913) with being "the first man to make a success of farming under the no-fence law by taking up trespassing stock under a law passed by the legislature in 1875." This was probably on his homestead on the North Tule. Wiley Crook and his two sons, Alexander and David, ranged cattle in the North Tule area starting in 1874.

Objections. The changes made in the cover and soil of the mountains by this long period of unrestricted grazing will probably never be known. Stockmen generally claim that reports of damage caused by livestock were greatly exaggerated. Nevertheless, some statements by early observers on the other side of the fence are of interest.

The geologist Clarence King (1902) says that the Kern Plateau was "green and lovely" when he first saw it in July 1864 (incidently one of the worst drouth years). But in September 1873 it was "a gray sea of rolling granite ridges no longer velveted with meadows and upland grasses. The indefatigable shepherds have camped everywhere leaving hardly a spear of grass behind them." King's observations illustrate a viewpoint that gained strength until it became an important factor in bringing an end to sheep-grazing in the Sierra.

John Muir had a big part in arousing the public against this sheep-grazing. With reference to the Tule River redwood forests after his hike through that area in 1875 he wrote (1901), "all the basin was swept by swarms of hoofed locusts, the southern part over and over again, until not a leaf within reach was left on the thorniest chaparral beds, or even on the young conifers, which unless under stress of dire famine sheep never touch." He had to quit his exploration of the Rogers Camp or Black Mountain grove of redwoods because the sheep had left no feed for his burro. On this same trip he wrote in his journal, "Nine-tenths of the whole surface of the Sierra has been swept by this scourge. It demands legislative interference" (Muir, 1938).

By 1900 the voices against sheep-grazing in the mountains were even stronger; for example, Dudley (1899) wrote, "From Nelson's Ranch I made four excursions. . . . but I found no space that had not been harrowed to dust by alien hoofs . . . . No one can imagine the destruction these creatures have wrought. . . . Probably most of this destruction had been worked by the nomadic Portuguese and Frenchmen. . . . Half a dozen forest fires were raging in sight as one stood on Jordan's Peak, above the old Jordan Trail on the 5th of September." This was the Camp Nelson area in 1898.

Surveyor P. M. Norboe (1903) wrote about two pine trees growing in the broken top of a sixty-foot Sequoia, and added, "They were about the only young trees in the forest that were absolutely safe from the all-devouring sheep."

In fairness to the "good" sheepmen it should be pointed out that most of the damage was done by "get-rich-quick" outfits that tried to beat the established sheepmen to the best ranges, thus grazing too early in the spring and forcing late-comers to over-graze or to use areas not suitable for grazing.

Landslides and Erosion. In December 1876, great landslides occurred in the Southern Sierra. Slides blocked the San Joaquin River and then let it go with a rush that wiped out Millerton, at that time the county seat of Fresno County. On the north side of Dennison Ridge a landslide scoured off the timber and four to twelve feet of soil, blocking the South Fork of the Kaweah and leaving a scar a half-mile wide and one-and-a-half miles long through the Garfield redwood grove (Taylor, 1960).

Kern and Little Kern lakes on the upper Kern River were formed by slides that year—not, as Doctor (1959) says, by the earthquake of 1872. Two of the first Tulare County sawmills were destroyed by similar landslides and floods on Mill Creek east of Fresno (Barton, 1907). The Kaweah and Kern slides

LANDSLIDES AND EROSION

contributed vast quantities of timber to the 1867-68 floods in Visalia and Bakersfield. People now living remember splitting posts years later from redwoods brought down by these floods on the Tule.

The surveyors who mapped the North Tule forest area in 1878 to 1883 reported several landslides, some of which are completely healed over now and barely distinguishable. There are also evidences of many small slides that probably date from 1867-68. In many cases they were stopped by redwood trees as evidenced by small benches around many large redwoods on the slopes facing the Wishon Fork near Shake Camp in Section 19 and 30. Old slides are also found in Section 11 south of State Forest headquarters, on north-facing slopes.

In passing, some mention should be made of a similarity between the events that preceded the severe erosion of the Southern California foothill ranges around 1860, and those preceding the severe mountainland erosion of 1867-68 in the Sierra. Cleland (1922, 1951) says that in the drouths of 1855-56 and 1860-61 cattle grazed everything they could reach and then died by the tens of thousands. This severe over-grazing was followed in 1861-62 by one of the longest rains on record, severe floods, and terrific erosion.

Don Jose Jesus Lopez, who was the major-domo of the Tejon Ranch in the 1850's, told Latta (1936) from his own personal knowledge that before this time there were no gullies—("barrancas") in the range country. He said that in the Mexican days the two-wheeled "carretas" traveled the bottoms of the dry ravines and they had no use for graded roads until these ravines were gullied out in the late 1850's and early '60's.

Heavy grazing in the high mountains came a little later, mostly after the rains of 1861-62. This was mainly sheep country. Paralleling the cattle situation in the previous decade, prices for sheep began to boom during the "Great Drouth" of 1862-64, and during these years the first heavy grazing of the mountains took place. Improper grazing or over-grazing of the high country by sheep can be very destructive to soil (Meeuwig, 1960). Then came the rains of December 1867. The unprecedented landslides and an unknown but undoubtedly significant amount of hillside, meadow, and stream-channel erosion occurred.

Again, the forest fires that probably swept the mountain areas during the early prospecting and grazing period may have contributed to the erosion potential. Everyone who has studied forest fires knows that a forest fire following a long period of non-burning can be very destructive to both vegetation and soil. This was probably the situation in the early sixties.

The evidence is in no sense conclusive that the great landslides were the result of improper sheep-grazing, in the way that the gullies on the foothill ranges resulted from overgrazing by cattle. Nevertheless, the fact that our most spectacular mountain erosion followed the first great sheep boom—in the same way that disastrous erosion followed the first great cattle boom—need not be considered a mere coincidence.

<u>Grazing Leases</u>. Stockmen apparently did not have to pay anyone for the privilege of grazing the mountain areas until after the Sierra Forest Reserve was created in 1893. Jay Brown has some interesting correspondence between Clinton Brown and Huffman White's brother Harrison White, dated from 1899 to 1903. In these letters Harrison White, who signed himself "Forest Supervisor" for the U.S. Department of the Interior, told Brown he would carry out the law without favor even to his brother's stepson, and that Brown must show ownership or written leases to privately-owned lands before he could get a permit to take sheep across government lands. And he told Brown that his requests to use wide strips of land and take five days to go ten miles would not be approved; and that "you are being closely watched."

Brown got a written lease covering 2560 acres of what is now State Forest land from Louisa Greenewald dated April 7, 1898, for the sum of $128. This lease was renewed each year until 1902, the fee increasing to $150. On the basis of the leases the government granted Brown a permit for 4000 sheep to enter the forest reserve June 21, 1903. Brown's lease from Greenewald required that, "he will protect the growing timber .... and stop any other person from trespassing." The latter stipulation was an important consideration in other grazing leases in subsequent years.

For more than forty years the sheepmen and cattlemen used the mountains almost without restriction. What management these lands had was supplied by these men. They developed practices, including the use of fire, that served their purposes very well. It was not easy for them to take a less important part in the management of the area and see Johnny-come-lately foresters apply a different philosophy of forest and range management.

Forest Fires. The Visalia Weekly Delta of July 29, 1875 carried the news that, "Heavy fires are raging in the mountains east of here. . . . [giving the appearance at night] of immense lanterns suspended from the heavens." The early prospectors and stockmen were not much concerned about forest fires. Heston in his letter of August 7, 1855 says, "The mountains for a vast extent have been burning and are burning now. It was set on fire about 10 miles from here (Green Horn Gulch, now in Kern County), and now presents a grand sight at night." (Tulare County Historical Society, 1958)

John Muir, although he praised the beauty and utility of some fires (Muir, 1901), made quite a point of their destructiveness, too, in the Sierra:

"Indians burn off underbrush to facilitate deer-hunting. Campers of all kinds often permit fires to run, so also the millmen, but the fires of 'sheepmen' probably form more than 90 percent of all destructive fires that sweep the woods--Incredible numbers of sheep are driven to the mountains every summer--and fires are set everywhere to burn off old logs and underbrush. These fires are far more universal and destructive than would be guessed. They sweep through nearly the entire forest belt from one extremity to the other." (Muir, 1876).

There is little evidence that Indians systematically burned the forests and much that they could not have done so (Burcham, 1959). Nowadays statements are often heard that the "natural" or normal condition of the Sierra forests was that brush was no problem, that a man could ride a horse anywhere and see considerable distances through the timber, and so on. It is interesting that some "old-timers" of a generation or two ago held exactly the opposite viewpoint.

Witness this quotation (Dudley, 1896): "The general testimony of the mountain and foothill people in regard to the changes that had occurred during the past 10 or 20 years in the vegetation of the mountains is not uninteresting. They assert that the undergrowth in the mountain forests has greatly decreased since sheep-herding came into the mountains. At present one can ride a horse anywhere through these high mountain forests, excepting in the inaccessible rocky places, while 20 years since it would have been almost impossible to have wandered far from the trails, on account of the underbrush, undoubtedly more dense than in the Northern Sierras. The sheep live on the young twigs of these undershrubs and on the small annual plants under the trees. The herders add to this destruction, as they pass out of the forest in the autumn, by setting fire to this undergrowth, in order to insure an abundant growth of tender sprouts in the spring following. The ranchmen believe this decrease in the undergrowth decreases the streamflow in the valley below during the summer, the water from the melting snows having little to hold it in check. They regard the destruction of the underbrush as more detrimental to the stream-flow than the destruction of the timber .... formerly many living springs were to be found on the ranches of the White River Valley, all of which now run dry in midsummer."

Apparently the "good old days" are the days of our youth, and they change with every succeeding generation.

It should be mentioned that Professor Dudley's statement was based on more than casual knowledge. Dudley was head of the Botany department at Stanford University. He and Frank Lamb spent July and August of 1895 zigzagging through the White River, western Kern, Tule, and Kaweah River forests from Glennville north to Sequoia National Park. He says, "Whenever we met a native rancher of the foothills, or the hunter and camper from the valleys, we discussed the utility and desirability of the reservations, and endeavored to get their point of view."

From various sources of information, including observations of the tree growth and reports from other Sierra redwood groves (Hartesveldt, 1962), the conclusion of this writer is that the period 1860 to 1890 was one of numerous fires both in the North Tule forest area and around its fringes. This was also the period of heavy use by the stockmen and lumbermen. The timber now growing on the Mountain Home State Forest shows that very hot fires occurred about this time, particularly in the upper Wishon Fork above Galena Creek and on the hills north and southwest of Old Mountain Home. Some of these fires apparently killed every live tree except a few of the large redwoods. Dense even-aged stands from seventy to a hundred years old are found now where those fires burned. Some fires were in logging slash (particularly those north of Frasier Mill) but most of them appear to have occurred a little earlier than the lumbering era.

Elliott (1883), writing about the Sierra redwoods of Tulare County, says, "Unfortunately, sheep men, in burning off the mountains to enable their sheep to penetrate the undergrowth, have destroyed hundreds of thousands of young trees of this species. . . . More than 90 per cent of all the young Sequoias germinated from seed in the forests of this county within the last 20 years have thus been destroyed by fires."

Deer. Another factor that probably affected tree reproduction was deer. Longhurst et al. (1912), in tracing the ups and downs of deer population in the Sierra, show that the deer population was very heavy before the hard winters of 1889-90 and the following decade, and very light during the nineties and the early twentieth century.

Surveyors' Descriptions. The descriptions that the surveyors of the U.S. General Land Office gave of the undergrowth and brush found in this area eighty years ago are of great value in judging vegetation changes. The surveyors of the Mountain Home area were definitely conscious of the small trees and brush. They wrote a description of the vegetation for each mile of line. The following is typical of W. H. Norway's notes: "North between Sections 35 and 36. Timber of pine, fir, cedar and redwood. Dense undergrowth of same, hazel, dogwood, and chaparral." Also "pine, fir, cedar and redwood. Dense undergrowth of same, manzanita and chapparrel," or "chemisal," or "chaparral and chincapin." Collins mentions "snowbrush" in addition to the others.

Baker found "Timber oak, elm, fir, redwood, cedar, and brush" on the line between Sections 2 and 11 south of State Forest headquarters and elsewhere throughout the township. Occasionally he found grass, for example: "Timber scattering oak and brush. Land rocky, produces good grass, soil third rate." Baker's "elms" were maples or alders. He wasn't a very good botanist but he probably knew brush when he saw it.

CHAPTER VII

FIRST SETTLERS, SAWMILLS, AND ROADS

> And as the din ariseth of wood cutters in the glades of a mountain. . . .   Homer, *The Iliad*.

WHEN PERMANENT SETTLERS began to arrive one of their first needs was lumber; and the transportation of lumber required roads.

First Settlers. The first settlers in the North Fork Valley, which borders our forest area on the west, came in 1863. They were the families of Jacob H. and Eleanor Cramer and Mrs. Cramer's parents, Cromwell and Priscilla Axe. (Could a Hollywood press agent have found better names for pioneers?) By 1869, when the Coburns came to the area, there were seven additional families and "old man Garner" established there. The families were those of Jim Flagg, S. N. Ellis, Jasper and Sam Webb, and Dave, Alex, and Wiley Crook (Stiner, 1956). By 1884 a number of changes and additions to the community had been made as is partially shown on the map.

Some settlers obtained free land under the Homestead Act of 1862; others bought 160 acres for $1.25 per acre as "pre-emption" claims (Act of September 4, 1841). The Cramers and Axes bought railroad land, probably because they had already used their homestead and pre-emption rights elsewhere. The Southern Pacific Railroad had acquired some of the better land in this area as a part of its government grant. The Cramer family settled on the North Fork and soon opened a store, postoffice, and wayside hotel. The Axes settled just east of Springville on what later became the Sequoia Stock Ranch.

An amusing story is told around Springville about how the nearest mountain, Hatchet Peak, got its name. In the early days of Tule River settlement, the two high hills that guard the lower entrance to the North Fork Valley were called Lumereau Mountain and Axe Mountain, for pioneers Charles Lumereau

FIRST SAWMILL 47

and Cromwell Axe. But one day Lumereau and Axe had a "falling out" over some cattle-grazing matter. It wouldn't have been neighborly for Charlie to take personal or legal action against Cromwell, but he felt it within his rights to rib him a little by cutting Axe Mountain down to Hatchet Peak. Charlie's private joke became public geography.*

First Sawmill. Three sawmills and one known shingle mill operated commercially in the North Tule forest area before 1884, and split products (shakes, posts, and rails) were cut in considerable quantity.

James R. Hubbs has been reported to have been the first man to manufacture lumber by means of a sawmill on the Tule (Stiner, 1934). This was about 1865. It was a water-power mill on the south side of the North Tule River above Jack Flats, in Section 9. A recorded deed from Asa Wetherbee to Benjamin B. Wetherbee dated November 16, 1869 establishes the name of this mill as the "Hubbs and Wetherbee Saw Mill." Stiner says it was operated first by Cramer, and this is possible. These old boys had to be jacks of all trades. The ditches that brought water to that mill are in good condition today, as are parts of the trestle-work that carried water to the top of an enormous overshot waterwheel. The vertical supports, some as long as twenty feet, are of sugar pine and accurately fitted by the mortise and tennon method.

Hubbs was the first of a long line of North Tule operators who probably spent more money than they made in the sawmill business. He is quoted (Stiner, 1934) as saying that it cost him $10,000 to put up the mill and build a road to it. Nathan P. Dillon bought this water mill and converted it to steam power.

A shingle mill is shown on the 1883 survey maps in Section 17. This was probably put in after 1875 by Dillon, or with money loaned by him, to utilize broken chunks from redwood trees cut for lumber or posts.

"In 1875 the lumber business reached its zenith," says Elliott (1883). In that year, according to Irvy Elster, Dillon leased the mill to Alonzo Elster (Irvy's grandfather) and Robert "John Doe" Harrington. They moved the mill higher up into

---
*Like Sir Walter Scott's Last Minstrel,
 "I cannot tell how the truth may be;
  I say the tale as 'twas said to me."

Section 10, and built the unique "car-track" or wooden railroad shown on the map. This track with 4 x 4 wooden rails replaced, for lumber hauling, the almost impassable road down to Dillon's Ranch. Alonzo Elster had been a railroad man in Ohio and had probably learned about steam and rails there. The cars loaded with lumber were let down the car-track by gravity and the cars were pulled back up by mules, according to the late Mrs. Emma Dillon Green of Lindsay, who drove the mules herself about 1877 when a girl of sixteen or less. These mules have been immortalized by the campground now known as Jack Flats and by nearby Jenny Creek.

The Dillon Mill was taken over, probably by lease, by Rass James (brother of one of James Hubbs' many sons-in-law) and P. Wagy about 1882. Wagy was a well-known sawmill man from northern Tulare County, described by Barton (1907) as one "who considered himself equal to any task and held himself in readiness to carry forward to a successful termination anything in which everybody else had failed." But he apparently had no better success than his predecessors. He later became County Treasurer.

This accounts for two of the early sawmills. The third—second in point of time—was the Wilson-Kincaid mill to be discussed later.

First Roads. The first wheels to roll into the North Fork of the Tule were probably those of the "small mountain howitzer" brought in 1856 to cannonade the Indians out of their Battle Mountain fortifications. However, there was no real road into the area until about 1863, when the first settlers came. According to a petition signed in 1877 by Huffman White and D. Gibbons, a road from Porterville as far as the Kincaid Ranch, "at the junction of roads to Dillon's Mill and Kincaid's Mill," had been in use "at least 13 years." (The old Kincaid Ranch later became the Charles and Clemmie Gill ranch headquarters about four miles north of Springville.)

The first road into the North Fork to be "declared" a county road by the Board of Supervisors was the one through Yokohl Valley. The "viewers" appointed by the Supervisors, May 11, 1872, were Jas. A. Kincaid and N. P. Dillon, and the road was to run from the vicinity of Dillon's grist mill about three miles east of Visalia, "through Yokohl to the Tule River Pinery Road."

THE WILSON-KINCAID MILL

This road is shown as a county road on the Official County Map of 1876.

The Dillon Mill Road (also called the Tule River Pinery Road) was declared January 4, 1882, but probably had been built about 1864 as an extension of the Porterville-Kincaid Ranch Road, for the purpose of hauling the first sawmill into that area. In 1877 Dillon petitioned for his road to be declared a toll road. A detailed survey was made and is on file in the court house.

The Wilson-Kincaid Mill.

Charles F. Wilson brought the second sawmill into our area, the first to make lumber from the present Mountain Home State Forest. It was hauled by oxen from Santa Cruz County to some as yet unidentified place near Happy Camp in Section 27 on upper Rancheria Creek. This was about 1870 (Stiner, 1934).

James E. Kincaid bought Wilson's mill at auction in 1876 for $400, but sold it and his "mountain home" a year later to Rand and Haughton. Kincaid must have had more to do with this mill and the area around it than his one-year ownership would indicate, because the name that seems to have been applied locally to this area before it was called "Mountain Home" was "the Kincaid Mill Country" (Brown, 1923). His biography says he operated a sawmill for several years.

Rand and Haughton moved the mill, soon after they acquired it, to the "Dome Rock" location shown on the map. The mill was located astride the creek just above the high waterfall near the northwest corner of Section 26 within the present Mountain Home State Forest. Up the south side of Rancheria Creek from the millsite, still visible to anyone who will search for them, are the old skid roads to the mill and some of the stumps of trees that were probably cut for it about 1880. The outstanding scenic beauty of this redwood grove has not been noticeably affected by the logging that took place there.

Roads To The Kincaid Mill.

There were apparently two roads to this sawmill. The public land survey plat of 1883 shows only the one up the ridge between the two main forks of Rancheria Creek through the Luther Carl Place and past Pine Springs. Elster maintains, however, that Wilson took his mill into the mountains from the Dillon Mill road following roughly the route of the present Balch Park Road. The two routes joined about two miles west of the millsite.

The earliest written reference found relating to either road is a petition on file at the County Courthouse which was presented July 20, 1870. It asked for a franchise to make a toll road from one mile east of "Sycamore Camp or Spring on the present Tule River Pinery Road, thence through what is known as the South Pinery in an easterly direction to a point on the Denison or Coso Trail."

The "Tule River Pinery" is now called Dillonwood and the "South Pinery" was the name then used for the area later to be known as the Mountain Home country. The "point on the Denison or Coso Trail" was most likely at or somewhere near the present Centennial Stump, because evidence on the ground and in various written and verbal statements indicates that there was a road to that stump before 1880.

Probably this road as far as the mill was built in 1872. The Visalia Weekly Delta of October 17, 1872 states that thirty men were "currently employed." The newspaper stated that the road would be used to bring lumber and fencing material from the mountains at reduced prices and also (October 24 issue) that a Mr. Wheaton was going to bring down and store ample supplies of that rare luxury, ice, and sell it for two cents per pound.

This road was in use at least until 1884, the year Coburn moved the sawmill out. It never was declared a county road. At least one part, the private road leading into Luther Carl's ranch house from the Balch Park Road, is still in use. Other sections are easily found on the ridge formerly called Summit Hill (Elliott, 1883) northwest of Churchill's.

Coburn's First Sawmill. Avon M. Coburn bought Rand's interest in this mill in 1883 or '84, reportedly borrowing the money from Haughton at eighteen per cent interest. It was apparently moved twice between the land survey of May 1883 and the making of the Official County Map approved December 4, 1884. The first of these moves was to a site still identifiable just across the road from Churchill's present cabin, and the second to the location shown on the map in this book. This latter site was at the Frank Knowles cabin, and the move probably followed close on the heels of the crew building the Bear Creek Road.

THE BEAR CREEK ROAD

The Bear Creek Road. This was the first road to reach the redwood timber in the Bear Creek drainage-area. It was requested in November 1883 by thirty-nine petitioners, headed by W. F. Vivian, J. R. Hubbs, I. McCutcheon, E. W. Haughton, and A. M. Coburn. Some other legible signatures were those of Frank Knowles, William Dunn, Martin L. Vivian (the man who cut the Grant Grove Centennial Tree in 1875), A. P. Osborn, C. J. Elster, J. E. McDonald, and C. M. Lumereau.

The road was to commence in Section 12 on or near the present Milo Road and "following as nearly as practicable the course of the Jordan Trail to the crossing of Bear Creek. . . . and up the Middle Fork, then crossing the township line. . . . at a point one mile west of the Twp. corner and from there to the summit of the divide between Bear Creek and Middle Tuly. This route crosses the lands of Jas. McDonald, William Dunn, E. W. Haughton, and Avon Coburn, the rest of the way being over public land." This describes the present road to Coburn's upper millsite a mile north of Knowles Cabin. The road was declared to be necessary because of "almost universal public demand. . . . the old pinery being completely exhausted."

J. M. McKiearnan was appointed "viewer" and on January 9, 1884, he and a Mr. Bradley (probably Abel Bradley, owner of land in Section 2 above the Knowles cabin) were awarded a contract to build it for $3,000. The maximum grade was to be "28 inches to the rod" and the road was to be eight feet wide. It was declared a county road August 12, 1884, but the official map of 1884 shows it ending at Frank Knowles' cabin, at least a mile short of the "Middle Tuly" divide. Elster says it was built by oxen, plows, and a "vee," and that his father, Charles Elster, drove the bullteam. The primary purpose of this road, of course, was to get Coburn's mill to the timber and to get the lumber out. The Visalia Weekly Delta of April 9, 1885, stated that "J. J. Doyle will continue the road on to his cabin at summit meadow." (Doyle's cabin was a hollow log in what is now Balch Park.)

This brings the road and trail situation to the end of 1884. The famous "Frasier Grade" was not built until 1885 (see Chapter XI).

The Lumber Business. This gives us an outline through the year 1884 of what were and would long continue to be the two main industries of our area, namely, the Dillon and the Coburn sawmills. Both mills during these years were small. Probably neither produced more than 20,000 board feet in any one day nor more than a half a million per year. Only five million board feet of lumber were sold in Tulare County in 1874 from six or seven mills according to the Visalia Weekly Delta of June 24, 1875, but a need for fifteen million was predicted for 1875. Most of the lumber sawed was pine. Redwood beams for foundation timber were probably made and a small amount of redwood boards.

Clear lumber was being advertised for sale in the Visalia Weekly Delta in July 1874 at $20 per M "all kinds," and common grades at $13 at the mill. Retail prices at Visalia were "clear lumber (extra) $45; redwood $42.50; common pine and redwood $33."

Among the many men who probably worked in these mills before 1884, Charles J. Elster (Alonzo's son), Jim Akin, Dave Wortman (father of Exeter lumberman Earl Wortman), and James McDonald are the only ones known, other than the various owners. A Mrs. Griswold (later Mrs. Moyle) cooked for the Rand-Haughton mill crew, according to Art Griswold, her grandson.

According to Joe McDonald of Springville, his father, James, was a carpenter and something of a millwright. He helped build the Rand-Haughton mill about 1877. Joe says that after his summer work at the mill he built a wheelbarrow and wheeled his belongings and his two children while he and his wife walked down the mountain to their new homestead they had purchased on Sycamore Creek just below the present Cypert or Greer ranch.

Some of the sawmills and shake-and-post cutters of this period must have operated on public domain timber without benefit of any purchase procedure. Most of the area was not surveyed until 1883, but even on surveyed land in California, there was no practical way for anyone to buy public timber or timberland until the passage of the Timber and Stone Act in 1878.

Biographical. Something should be said about the men, heretofore only mentioned, who pioneered in lumbering in this area. The people who settled the North Tule and carried on the forest activities there were almost exclusively Americans of Anglo-Saxon stock and from the northern states. No Spanish influence

JAMES R. HUBBS

was in evidence. In fact, the community still has not fully accepted the Spanish ingredient that flavors the speech of most Californians. Springville's annual celebration is still the "Road'-e-o"; not "Ro-day'-o." Most come from forested states and had experience in woods work.

James R. Hubbs. James R. Hubbs was born in Illinois. He and his first wife, Eliza Farrell, came from Arkansas to Visalia in 1853 by ox-team (Stiner, 1956). In 1879 he moved to the present Springville area, operating a cattle ranch from headquarters located where the U.S. Forest Service Ranger Station is now. He was a leader in many activities in Tulare County. He and N. P. Dillon were mentioned in a state-wide list of "Landowners owning 500 acres or more in 1872," published in the Sacramento Daily Union, February 1, 1873.

Hubbs' first experience in sawmilling seems to have been his last, but his son-in-law, Charles J. Elster, and other relatives stayed in the game. Two of his daughters by his second wife, Mary Jane Dunn Hudson, married Charles Elster--in tandem, of course; not two abreast. Eva, the eldest, was Charles' first wife; Minnie, the youngest, his second.

Nathan P. Dillon. "Nate" Dillon was known in early-day Tulare County as a large land-owner, a gristmill operator, and somewhat of a philanthropist. The "Tule River Pinery" on the North Fork, however, was one of his greatest and most lasting interests. This property consisted of about 1,000 acres of pine and redwood timberland, a sawmill, and a shingle mill. He first appears in the record as a "druggist and chemist," proprietor of "The Old Drug House" at the corner of Main and Church streets in Visalia, according to advertisements in the Tulare County Times of December 26, 1874 and succeeding issues.

Dillon was probably born in Pennsylvania but he came to California in 1852 by way of Illinois, the Mormon settlements in Utah, and the California mining districts. He was disillusioned by both Mormonism and the mining towns, and settled at or near Visalia. He was thirty-two years old then. His youngest daughter, Mrs. Veda D. McCoy, says that he worked at lumbering in Illinois. His wife's people, the Van Leuven family, came across the plains with the Dillons but settled in

San Bernardino and have become a numerous and well-known family.

According to Clyde Osborn, who now owns the old Dillon homestead, Dillon took up land in 1868 in Section 17 on the North Fork of the Tule and developed irrigation for an orchard. He did not move there until several years later. (Courthouse records indicate that Dillon's patent was not entered until January 15, 1896. Such time lags between homesteading and patenting were common.)

He and his first wife, Zilpha, had twelve children. She died in 1888. When he was about seventy he married Nellie L. Marshall (a niece of James Marshall, the discoverer of gold in California) who bore him a daughter, Veda. Nellie Dillon was killed in a team runaway. Nathan died in 1903 leaving an estate worth $100,000, according to the Lindsay Gazette. His timber property was disposed of in 1927 for only $8,000. A burial vault in the Porterville cemetery houses the remains of Dillon and his two wives.

Among the children of Nathan and Zilpha were Leander, George (who homesteaded the battleground of the "Tule Indian War" and was reported to be one of the best teamsters of his day), Delbert (who homesteaded Upper Grouse Valley), Alma, Henry, Ellen Kinkade, Amanda Johnson, Emma Green, and Anna.

There are many stories told about how Dillon acted as banker for his neighbors. He helped finance many small lumber operations, both on his own property and elsewhere. His daughter, Mrs. McCoy, says:

"My father became blind when I was three years old. He taught me about numerals and coins as soon as I was old enough. I was his eyes when he loaned money. He always carried a buckskin bag in his left pocket with gold coins in it. He carried silver coins in his right pocket. The amount would usually be around three or four hundred dollars. When a neighbor came to borrow money from him he would call me to help him count it. He would take the gold pieces and feel them over for size, then around the edges. He would pass them to me for verification, then to the neighbor without a note or signature. When the money was returned the same procedure took place only vice versa. He could figure compound interest in his head faster than the average person can with a pencil."

Dillon stood out as an individual, even in the days of rugged individualism. His refusal to follow the crowd is illustrated by the following incident from Menefee and Dodge (1913), who quote Stephen Barton's account of 1874:

"Nathan Dillon, Wiley Watson, Mr. Kenney and others, feeling that it was an outrage to drive Indians to the wall on so slight a pretext, undertook to remonstrate. These men were among the most high-minded and substantial citizens of the county but their arguments proved without avail. The tribe camped a mile below Visalia were ordered to surrender their arrows and move their camp to the western edge of the town." [After the failure of the whites' first attack at Battle Mountain], "Having ingloriously fled from the field of battle, this force now sought a cheap plan of retrieving a reputation for heroism by turning on those citizens who had counseled moderation and fair dealing. . . . [They] required that those who opposed the war should, at the peril of their own lives, as well as the lives of the Indians involved, convey the Indians out of the settlement. . . . Dillon gave $10 and a thousand pounds of flour. . . ., and Jim Bell was hired to take a heavy ox-team and haul the poor outcasts to Kings River."

Charles F. Wilson and James A. Kincaid. These men both came from Pennsylvania by way of Michigan, and both came to Tulare County about the same time but from different directions. Kincaid's background might be considered typical of many Mountain Home pioneers. He "spent the early years of his life working in the woods of Pennsylvania and Wisconsin, cutting and hauling logs and running them down the rivers to the mills in which he worked" (Stiner, 1934). After being lamed by a kick from an ox he was educated as a teacher but overcame his physical handicap enough to farm in Minnesota, pre-empt a ranch in Frazier Valley in 1871, and engage in the forest activities already mentioned.

Avon M. Coburn. Avon M. Coburn, the latest to appear on the scene as a mill owner, did not come into the business unprepared. His father, Samuel S. Coburn, came from the lumbering state of Maine and, among other occupations, was a blacksmith. He came to California in 1850, and in 1869 located five miles north of the site of Springville. He was famous for being able to lift oxen and other things, including the flowing bowl. His son, on the other hand, was a temperate man. Avon was born in Placer County in 1857, but was sent to Maine to go to school until he was fifteen. He then joined his father in the North Tule and worked in sawmills from 1872 to 1884 (Quinn, 1905), at least part of the time at the Rand and Haughton Mill.

Coburn was somewhat of a mechanical genius. He is credited with having built the first motor boat in the area and the first telephone system, the latter having been built "from scratch" using specifications obtained from the U.S. Patent Office. He used it only as a private signal line between the two ends of his lumber flume, thereby avoiding suit for infringement of the patent (see Chapter XI).

CHAPTER VIII

AN OLD STUMP AND TWO CAVES

> The Mountain Room is the highlight of the cavern . . . this single chamber comprizes the largest cavity in California and Nevada known to the Western Speleological Institute. . . . The feeble carbide lamps seem like mere candles in their futile attempt to illuminate this chamber . . . like constellations in a storm-clad sky, colonies of brilliant white draperies and stalactites catch the illumination of the lamps. . . . Other features include gours and rimstone, refulgent stalactites and stalagmites, stalactite ribbons formed along ceiling joints, cave pearls, flowstone cascades, to mention but a few. . . . For delicacy and grace, the drapery stands out as the foremost cave feature ever encountered by the members of the Survey. The unviolated preservation of this drapery alone merits whatever expense may be necessary to protect Haughton's Cave.
> Raymond de Saussure, "Preliminary Exploration of Haughton's Cave" (7-15-1952).

The Centennial Stump. The tree that once stood on what is now called the Centennial Stump was the largest tree of the area and one of the first to be cut down. Almost everything else written or said about this tree is disputed or unverified. In 1876 the nation celebrated the one hundredth year of its independence by holding the Philadelphia Centennial Exposition. This event was written up in magazines and newspapers for several years in advance, giving time for developing the idea of exhibiting a giant redwood there to advertise the new State of California and to make a profit doing it.

There are at least two Centennial Stumps in the Sierra. The one near Grant Grove in Fresno County is well documented by newspaper accounts (Tulare County Times, March 27, 1875; Visalia Weekly Delta, April 1, July 15, September 23, and November 18, 1875) which report its travels as far as St. Louis.

The Mountain Home tree, on the other hand, apparently received no contemporary newspaper notices at all. The published information about both stumps has been well reviewed by the Tulare County Historical Society (1950). In many local stories the Centennial Tree is confused with exhibition trees cut later at Mountain Home, viz., the McKiearnan tree cut in 1889, the World's Fair tree of 1892, and the Nero tree of 1903 (see Chapter XII).

The earliest mention found of the Mountain Home stump as a "centennial stump" is in W. H. Norway's public land survey notes of May 14, 1883 as follows: "East on a random line between Sections 25, 36. . . . 23.19 chains, a house at centennial stump bears North 25.00 chains distant." The Mountain Home stump was the result of a venture by three Tulare County men, James R. Hubbs, John McKiearnan and Ed Manley (Elliott, 1883; Brown, 1923; Tulare County Historical Society, 1950). Bill Dunn and Jim and Henry Talley did the axe and adze work, according to Elster.

Mrs. Brown's Story. Mrs. Jay Brown (1923) of Porterville made an heroic effort to gather the facts about the Centennial Stump. Since her account of the cutting of this tree is not available to everyone, parts of it are quoted below:

"The men started erecting a scaffold about 20 feet high, around the tree they had chosen. The tree measured 111 feet in circumference at the ground and 26 feet in diameter at the point where the cut was made. By a careful count of the rings at the time the cut was made it was estimated that the tree was well over 3,000 years old.*

"The scaffolding built, the men felled the tree, leaving the stump about 24 feet high. From the top of the tree, which was about 200 feet in length and 25 feet in diameter at the butt, after the bark was removed, several thousand fence posts were cut. . . . Working from the top of the stump they had left and leaving a rim of wood from six to seven inches thick, just enough to show the red inner wood against the lighter colored sap and the distinctive fibrous bark, from three to twelve inches thick, they chopped away the heart of the stump to a depth of about sixteen feet leaving the stump to that point in the form

---

*The Douglass research in 1918 proved this count to be very accurate.

of a barrel.* This accomplished, they began again at the top of the rim, ripped the hollow shell lengthwise with whipsaws into 15 equal sections and cut the slabs off about 10 feet from the ground on the lower hillside and about four feet from the ground on the upper hillside. The slabs were lowered by means of rudely constructed derricks and cables and numbered in rotation ready to be hauled out as soon as the roads opened up in the spring of 1876. The men spent about a year working on the tree.

"In June, 1876, the slabs were loaded on wagons, each stave making a load for a six-horse team, and were hauled over steep mountain roads, two miles of which had to be constructed especially for this purpose over what had been merely a trail, to Mountain View, north of Springville. . . . The road followed the old Tuohy trail to Kincaid's Mill, through Happy Camp and Pine Springs and Rancheree roads to Mountain View. Here, about July 1, 1876, the slabs were set up, fastened together from the inside with iron hinges, hand forged by local blacksmiths, with a small opening left between two of the slabs so that the inside might be inspected. . . .

"On July Fourth, 1876, Independence Day was celebrated at Mountain View in true mountaineer fashion, crowds of people coming from Springville [Note: There was no Springville in 1876.] and Porterville, then only a small sheep town with no railroads. When a few days after the celebration the tree was again taken down and hauled over weary miles of dusty roads, through Porterville to Visalia where it was exhibited a short time on the lot where the Visalia Theater now stands. From there it was hauled to Tulare. . . .

"Before leaving Mountain View, Mr. Manley disposed of his interest to Messrs. McKiearnan and Hubbs, who shipped the tree from Tulare to San Francisco where it was set up, this time in Woodwards Gardens at Thirteenth and Valencia Streets. Admissions were charged and the curiosity gained considerable notoriety, even among Californians, few of those about the bay at that time having visited among the Sequoias. Early in August Mr. McKiearnan sold his half interest to Mr. Hubbs, taking in payment 14 head of cows, and returned to his home at Springville. Mr. Hubbs remained with the exhibit a few days longer

---
*It is hard for us these days to imagine digging a well 24 feet in diameter in solid wood across the grain with only axes and adzes to work with. But then, perhaps as in the Biblical Land of Canaan, "There were giants in those days"—men to match the giant trees.

and then sold the tree for two thousand dollars which was three thousand dollars less than the undertaking had cost its promotors."

Mrs. Brown's account continues with descriptions of the showing of the tree in San Francisco "in August 1876" (which seems too late in the year to attempt to show it also at Philadelphia) and at the Centennial in Philadelphia, all of which is doubtful in the light of later research. She includes Orien McKiearnan, Jim Hubbs, Jr., and William Dunn among her sources of information. These men should have had the facts about the local events.

The road by which the tree was hauled out is practically all abandoned. The route can still be retraced, however, by sections of hand-laid rock wall, remains of a log "corduroy road" in a wet place, etc., even though parts of it probably have not been used for seventy-five years.

Woodman, Spare That Tree! Public opinion at the time of the cutting of these trees was divided. The Tulare County Times (March 27, 1875) called Martin Vivian's project near General Grant an "outrage." Elliott (1883) says he should have been imprisoned for life for his "vandalism." (He plead guilty in a state court before undertaking the job and paid "the highest fine the law imposes," according to the Delta of April 1, 1875.) The Delta defended him on the grounds of his age (seventy years), the advertisement he would give the state, and because he would spend $8,000 or $10,000 on the project. This paper said the U.S. General Land Office was being besieged by protesting people but the record is not clear as to whether he was ever brought into a federal court. (Perhaps here is a clue as to why the cutting of the Mountain Home "Centennial Tree" was not reported by the papers.)

Date of Cutting Uncertain. As to the date of cutting, there is wide disagreement. Douglass (1945-46) writes that his guide, Charles Elster, told him he helped cut it in the winter of 1874-75. Mrs. Brown gives 1875 as the year. Elliott's account is somewhat jumbled as between the two "Centennial Trees," but we may assume that his 1878 date applied to the Mountain Home tree. It is far from a definite statement of the year of cutting, however. Later histories apparently based

HAUGHTON CAVE 61

their stories on Elliott's. A photograph in Mrs. Gertrude Oldham's album is inscribed "Cut in 1877."

In an attempt to verify Mrs. Brown's account of the exhibition of this tree at a centennial Fourth-of-July celebration at Mountain View schoolhouse in 1876, an announcement was found (Delta, June 15, 1876) that such a celebration was to be held, that James A. Kincaid would read the Declaration of Independence and Mr. Seaman (probably Joshua Seamonds, one of the first landowners of present Mountain Home State Forest land, and probably also the "Seaman" who accompanied King and Knowles to Mt. Whitney in 1873) would be marshall for the grand dance, but no mention of the exhibition of any tree.

Douglass probably could have accurately dated the last ring on this stump when he saw it in 1918. He identified and dated every ring back to 1122 years before Christ was born, but does not mention the year of the outside ring.

Negative evidence as to 1875 as the year of cutting is that John Muir in his reports (1901, 1894) of his trip through the area in 1875 does not mention the cutting of any tree, and the Official Map of Tulare County by P. Y. Baker, dated August 1876 shows, at the approximate location of this tree, a drawing of a large redwood with the notation, "Largest tree in world, diameter 46 feet." One possibility as to the origin of the name "Centennial Stump" is that it was first called the "Centennial Tree" because of the publicity given it by its appearance on this 1876 map.

To sum up, it appears that although the connection between the Mountain Home Centennial Stump and the Philadelphia Centennial Exposition is entirely unsubstantiated, there must have been some connection in people's minds at that time between the tree and the hundredth year of America's independence. It seems most likely that the tree was cut in 1877 and hauled out in 1878, and that the name "Centennial" was attached to the tree because of an intention to cut it for the Centennial Exposition but the cutting had to be postponed. Perhaps the original plan was abandoned because Martin Vivian had gained a head start. There seems to be no evidence that it reached the grounds of any of the great exhibitions of that period.

Haughton Cave. E. W. Haughton did not mingle very much socially, but he did get around in the woods. And one day, probably while looking for a good piece of redwood timberland, he found a big limestone cave in a sharp little gorge at the foot of a high cliff. He almost failed to come back from that trip and people remembered his story so well that they called it Haughton's Cave.

Haughton's story, as remembered by Elster, is that he went down into the cave (it can be entered only by an almost vertical passage) with only pitch-wood torches for light. Between the smoke of the torch which almost suffocated him and the darkness without it, he felt fortunate to get out alive. This was in 1884 or before, because the Official Tulare County Map of that year shows the cave with Haughton's name. The map location is about three-fourths of a mile southeast of the actual location as known today. In the nineties, when tourists began to flock to Mountain Home, this cave was given quite a bit of publicity (Lewis, 1892). Jesse Hoskins regularly took people down into the cave by means of fir-tree ladders made by cutting trees and leaving the branches "trimmed long" for steps (Keagle, 1946). The cave was usually called Crystal Cave during this period.

Haughton was a bachelor. Mrs. Hubbs says he was English, but an article about him in the Porterville Farm Tribune of November 2, 1950, says that he was born in Ireland, came to California in 1849, and served with Jim Savage's "Mariposa Rangers" on their 1850-51 expedition to Yosemite Valley to punish the Yosemite Indians. This article contains a detailed account by him of this expedition. The style of the article indicates that he was an educated man.

A clipping in Stewart's scrapbook (1933, p. 72) from the San Francisco Chronicle of September 9, 1900 shows that Haughton had some claim to fame among early Californians: "E. W. Houghton, one of the discoverers of Yosemite, was found dead in his cabin in the Sierra east of Tulare. Houghton came to California in 1849 and was said to be the first white man to see Yosemite. He knew every Indian trail from Mariposa to Kern River." (The misspelling of his name is in accord with the "Roster of the Mariposa Battalion" in Robert Eccleston's The Mariposa Indian War.)

Haughton's name is variously spelled. The Farm Tribune article gives it as Edwin W. Houghton but court house records usually show it Edw. or E. W. Haughton. His contemporaries apparently pronounced it "Horton" and often spelled it that way. He settled on what is now called the "Bear Creek Ranch" in Section 9, probably before the road was built in 1884. By the end of that year he had also acquired as a "timber and stone claim" the east quarter of Section 35 including the sites of Frasier's Mill and Coburn's upper mill on the Coburn Fork of Bear Creek. The following year he acquired land in Section 36 from the State. His sawmill partnerships have already been mentioned.

He is said to have been a good friend of Frank Knowles, but not of his other neighbor, John Gaffney, an Irishman, who, with his sister Ann, ran a half-way house or hotel at Rancheria. And Gaffney's hogs coming the three miles up the road and getting into Haughton's garden didn't help neighborly relations. Haughton used to shoot the hogs. Mrs. Ola Hubbs tells the story that one day in 1900 Avon Coburn found that Haughton had died at his well-kept bachelor cabin, and he stopped at Gaffney's on his way to town and told him the news. John's reaction was, "Sure, and he's probably in Hell by now, a-yellin' and shootin' at my pigs." The <u>Farm Tribune</u> article says he left land holdings worth $20,000 to $25,000, and that he was buried in the Porterville Cemetery.

Haughton's place has often been referred to as "the Coburn dump," because it was the lumber drying yard at the lower end of Coburn's flume. After Haughton's death, it was found that he had not actually owned the land where his improvements were. Clayton Northrop therefore was able to acquire it under the Forest Homestead Law. Two of Northrop's children were buried there in 1915, in the cemetery site with the large natural headstone visible from the road. Pete Planchon (father of Bill Planchon, Northrop's son-in-law), and Jake Garner were later owners of this ranch, and it is now operated by the Ratzlaff family as a productive apple orchard.

<u>Galena Cave</u>. There are other caves in the Mountain Home area, but none of them are definitely known to have been entered. An undated clipping from a Visalia paper, filed in the Tulare County Library, tells about Alf Powell running a prospecting tunnel "into a labyrinth of limestone caves" in that area and taking out some ore that appeared silver-bearing. Elster says this cave was called the Galena Cave. The cave described as follows by Mr. W. L. Nunes of Fresno, in a letter of September 11, 1960 to the writer, is probably the same one:

"I submit the following information recently related to me by my father, John J. Nunes, 223 North "E" street, Porterville. My father, as named above, together with William Grider of Hanford, was employed by four Porterville businessmen (J. H. James, Frank Woodley, Eugene Scott and Henry Traeger) to extract galena ore from a deposit in the Mountain Home area in the late summer of 1901.

"The site of this operation is described as adjacent to one of the creeks on the east side of the North Fork of the Middle Fork of the Tule River, in the gorge below and almost directly east of the then existent but inoperative Enterprise Mill.

"Transportation to the mining area was accomplished from Porterville by horse-drawn wagon to the Enterprise Mill and thence by foot into the canyon. Provisions at intervals were transported via burro from this same mill and the animal used was the property of the Powell Brothers (Al and Monty) whose copper mining operations a short distance down-stream on the Tule River were supplied in the same manner.

"The close cooperation between the copper mining enterprise and the adjacent one in which my father was involved indicates that the galena claim may have been discovered by the Powells who had arranged with others for its working. At any rate, the brothers were occasional visitors to the upper mine during the six-week period of my father's employment.

"In the mining; . . . . a vertical shaft approximately four feet square was sunk into the mountainside.

"The ore and waste material was raised from the bottom of the shaft in a bucket by means of a hand-operated windlass turned by one member of the team while the other performed the excavation process below. When a depth of 12 or 14 feet was reached, Grider, digging in the shaft, was understandably startled when most of the bottom of the excavation fell away into a void below, leaving intact only a small ledge upon which he was standing. It appeared that the shaft had broken into a limestone cavern of great depth. A wooden slab from a nearby fallen tree was then wedged into the shaft to provide a platform from which the workman below might then remove ore from the side-walls of the excavation. Henceforth, all waste material was allowed to fall into the shaft and some indication of the magnitude of the cavern below was apparent by the fact that rocks loosed into the opening could be heard for several minutes making their descent."

The location of the above operation has recently been rediscovered but the shaft has collapsed and has not been reopened.

PART TWO

NOW WE'RE LOGGIN'

> And Solomon sent to Hiram saying, command thou that they cut me cedar trees out of Lebanon; . . . for thou knowest that there is not among us any that know how to cut timber like unto the Sidonians. . . . And Hiram sent to Solomon saying, . . . I will do all thy desire concerning timber of cedar and timber of fir. . . . And Solomon had threescore and ten thousand that bare burdens, and fourscore thousand that were hewers in the mountains.
> *Old Testament, I Kings; 5.*

THUS DID KING SOLOMON "gyppo" to Hiram and his Sidonian axmen the logging of the cedars of Lebanon. And a few thousand years later in America the lumber industry hewed its boisterous way from Bangor, Maine to Eureka, California with little basic change in philosophy or methods. But when men went up the Tule River to bring down the redwood and pine they did some things that were not traditional. For one thing they usually took their families along. Perhaps this is one reason that lumbering history in this area is so closely interwoven with the development of recreation activities.

Part One covered broadly the whole landscape of the North Tule forest area from Indian times to 1885. In the following five chapters the narrative is continued through the next twenty years, the heyday of private enterprise. The grazing story is not included because it was mainly a continuation of the preceding period. Sawmilling, recreation, and the beginning of the conservation movement in this area are the main subjects of Part Two.

CHAPTER IX

A WORD TO THE WISE

> Gain to the verge of the hogback ridge where the vision
>   ranges free:
> Pines and pines and the shadow of pines as far as the eye
>   can see;
> A steadfast legion of stalwart knights in dominant empery.
>   . . .
>
> Wind of the East, Wind of the West, wandering to and fro,
> Chant your songs in our topmost boughs, that the sons of men
>   may know
> The peerless pine was the first to come, and the pine will be
>   last to go!
>
>   Robert W. Service, *The Pines*.

THE U.S. CONGRESS passed the Timber and Stone Act on June 3, 1878. Twenty-five days later the <u>Visalia Weekly Delta</u> reprinted the Act. The newspaper called it "an act for the sale of timber lands" and added, "A word to the wise is sufficient." By October of that year surveyor P. Y. Baker was running a survey line north across the Wishon Fork near the present Camp Wishon headed toward the Mountain Home redwood area. By the end of 1883 all the land in the North Tule forest area had been surveyed except some precipitous slopes that have not yet been sectionized.

<u>The Timber and Stone Act</u> provided that any citizen could buy 160 acres of surveyed public land for $2.50 per acre with no required residence or improvements, if he would swear to occupy or use the land himself (Greeley, 1951). "Undoubtedly this was the most abused in practice and consequently the most criticized of all the actions of Congress in the public domain issue." (Clar, 1959)

It should be said, however, to the credit of the people of Tulare County that no land fraud charge involving lands of the

North Tule forest area has ever come to the knowledge of this writer. This seems quite remarkable in view of the way redwood and Douglas fir lands were acquired in the great forest regions of the West Coast, and even as near as Converse Basin in Fresno County (McGee, 1952).

Another way people could acquire surveyed timberland was to buy it from the State. Sections 16 and 36 of each township plus all lands classified as "swamp and overflow" passed to State ownership as soon as they were surveyed. The proceeds from their sale were to be used for school purposes.

In California "school lands" and "swamp lands" were sold as rapidly as buyers came forward with the legal price of $1.25 per acre. In fact, prior to 1885 much timberland was bought for twenty per cent down (25¢ per acre) and "the rest when you catch me" (Clar, 1959). The surveyors did not classify any land as "swamp and overflow" in the North Tule forest area, but school lands included two Sections 36 (one in the Mountain Home State Forest and the other above Dillonwood) and Section 16 above Camp Wishon in the Alder Creek redwood grove. Some properties outside our area but just east of it were classed as "swamp and overflow." Two of these are known as Peck's Cabin and Junction Meadow. More distant areas so classified included Giant Forest and Log Meadow in Sequoia National Park.

We can imagine that forward-looking people with a little money to invest followed very closely the progress of the public land surveyors in the North Tule forest area. The school and "swamp" lands were probably the first to be taken as the price was less, but it appears from county records that the best of the federal timber lands were filed on very soon after the plats were approved.

<u>The Surveyors.</u> Peter York Baker, the first government surveyor to enter the North Tule forest, was a prominent citizen of Tulare County, one of the founders of the city of Traver, and, in 1882, a County Supervisor (Lewis, 1892). He had been a county surveyor and clerk in Kansas. Baker surveyed the southern part of the Bear Creek area before the end of 1878. His notes and plats are very accurate considering the standards in existence at that time.

The next surveyor in the North Tule forest area was G. S. Collins who surveyed Townships 19 and 20 South, Range 31

East, in 1882. He apparently operated mostly from camps in the plateau area east of Maggie Mountain, because his work is reasonably accurate there. His surveying in the Wishon Fork drainage is very sketchy and some of it was surveyed only on paper. The fourth and last township in the area was surveyed in 1883 by W. H. Norway.

Private Ownership Reaches the Timber. The eighties was a time when nobody expected much from the government except easy access to the natural resources of the West and a free hand in developing or exploiting them. Until this time almost all of the mountain areas were still public domain. This suited the sheepmen well enough because they had been using it for thirty years without paying taxes or fees and had no objection to continuing the arrangement. For lumbering and the summer resort business, however, ownership was desirable and conditions were shaping up that would make it possible.

The year 1884 may be called the year of the "land rush" for timberland in the North Tule mountains. The story can be roughed out from the records of the General Land Office and Tulare County. J. D. Hyde, register of the Visalia Land Office, received and filed the township plat for the first of the four townships of the North Tule Forest area on February 28, 1879, but if this resulted in any activity in the forested area the records do not show it. But after Mr. Hyde filed the other three plats (two on September 14, 1883 and the other on February 9, 1884) he began to do a "land office business."

The earliest record found of a filing was on "school land." Mr. S. M. Gilliam (a Tulare County supervisor) filed a "location" on the north half of Section 36 on December 15, 1883, but apparently did not follow it up. J. J. Doyle, well-known later as a real estate man and owner of the Summer Home resort, filed on the same property March 10, 1884. A. M. Coburn, the Springville lumberman, filed on the south half of this section February 20, 1884, and received a patent in 1889.

On land patented directly from the federal government the earliest record found is a deed from Joshua Seamonds to Mary L. Seamonds dated April 21, 1884 for 160 acres in Section 26 on upper Bear Creek, even though Seamonds did not receive the patent from the government until 1887. In 1885 Mary deeded the property to the lumberman L. B. Frasier.

BLOCKING-UP TIMBERLAND

The map shows that at least twenty-three private citizens had laid claim to lands in the forested region by 1884, all in the Mountain Home area. The earliest date of a patent, however, seems to have been in the Dillonwood area. Nathan P. Dillon's 80 acres in Section 10 was patented January 15, 1885. In addition to the names shown on the map there were several others known to have taken up land in the same area about the same time. They include Mrs. Sarah M. Doty of the Mountain Home Resort in Section 35, W. T. F. Smith and John G. Eckles in Section 26, Elizabeth J. Shirley in Section 19 and 30, Coleman Talbot (father of Courtney and the original California immigrant of the Talbot clan) in Section 31, all in the Mountain Home area; and George Dillon in Section 3 and Clara A. Lindsay in Sections 3 and 10, Dillonwood area.

Four of the men whose names appear in the Mountain Home part of the map (Courtney Talbot, Wm. H. Hammond, S. M. Gilliam, and Wm. J. Newport) were county supervisors. Perhaps the most active families were the Talbots and the Doyles. Courtney Talbot and John J. Doyle married sisters, daughters of Conrad and Sarah Holser.

At least twelve of the entrymen in the Mountain Home area received patents dated the same day, July 25, 1887. All were "entered" from January to May 1884, and six were to J. J. Doyle and his known relatives. All but one was for "cash," from which it is assumed that they were Timber and Stone claims for which the entire $2.50 per acre was paid upon entry.

Most, if not all, of these people apparently "took up" this timberland in good faith under the Timber and Stone Act of 1878. If there were exceptions they were probably in the area now owned by the State in the upper Wishon Fork drainage-area in Sections 18, 19, 20, and 29. Statements have been made that J. J. Doyle and Smith Comstock about 1890 promoted the filing on lands in this area with the understanding that Comstock's Tule River Lumber Company would buy the claims.

<u>Blocking-up Timberland.</u> Not all of the original entrymen, no matter how sincere their intentions of making personal use of their claims, were able to do so. In some cases they were unable to pay the taxes and lost their properties to the State. One such tract was a 13-acre plot south of the Frasier Mill in Section 35, which was lost to the State in 1887 for non-payment of $14.66 taxes. The late Chester Doyle of Porterville stated

the Doyles cut 50,000 redwood posts after buying it from the State for $75. In other cases tracts were sold for rather fancy prices for those times. For instance, in 1886, according to county records, Allen McFadgen paid Amelia Talbot (Courtney's wife) $3000 for the 80 acres between Shake Camp and the site of the Enterprise Mill. This property under the name of "the McFadgen 80," became widely known for the fabulous stand of redwood that grew on it, reported at 8 million board feet (Walker, 1890). The stumps still standing on it make this estimate credible.

The first company to try to consolidate a large block of timberland was the Tule River Lumber Company, incorporated December 24, 1889 with a capital stock of $500,000. Austin D. Moore, Jacob Levi, and John R. Jarboe of San Francisco, and Hiram C. Smith and Smith Comstock of Tulare County were the incorporators, owners of all the stock, and the original directors. Smith Comstock had been well-known since 1879 as a successful sawmill operator in the Millwood and Atwell's Mill areas. The names of Smith and Moore as early as 1885 or '86 were connected with the large redwood lumbering enterprises at Millwood (Tulare County Historical Society, 1950). Since Comstock acquired some Mountain Home land in his own name before the company was incorporated, it is believed that he was the active member of the firm in our area.

When Thompson's Atlas of Tulare County (1892) was published, the Tule River Lumber Company was shown as the owner of about 2600 acres of Mountain Home timber, all within the present Mountain Home State Forest.

In Dillonwood, Nathan Dillon got together about 1000 acres and a Mr. Canty about 500. All privately-owned land in that area that they did not acquire passed later to what is now the Sequoia National Forest.

It appears from the county records that the Tule River Lumber Company immediately after its incorporation, mortgaged the property Comstock had bought up, to Louisa Greenewald of San Francisco for $60,000. In 1896 the Greenewalds brought suit to foreclose and received a deed April 7, 1897. The Tule River Lumber Company was suspended as a corporation in 1905.

In February 1904 Greenewald sold the property to T. C. Judkins of San Francisco. Judkins was a director of the Western Pacific Lumber Company which was incorporated

BLOCKING-UP TIMBERLAND 71

June 23, 1904 with F. A. Boole and Wm. G. Uridge of Fresno also on the board. More will be said later about Uridge. In this case also, Greenewald took a mortgage, this time for $36,000. Later that year the title passed to Uridge, in 1907 to Archibald Kains, back to Uridge, then to Ira B. Bennett, and eventually to the State of California (see Chapter XVI).

CHAPTER X

THE TREES COME DOWN

> Under conditions peculiar to the southern ranges of the Sierra Nevada Mountains in California, it is probable that the nation's most skilled teamsters were developed between the Gold Rush and the advent of the automobile. . . .
> What a beautiful sight it was to see an eight-horse team take a loaded wagon around a hairpin curve on a steep mountain road!
> Monroe C. Griggs, *Wheelers, Pointers, and Leaders.*

> It is a thrill to return to the days when the lumber teams came down Putnam and made the turn on Main Street, the swing teams pointing and the bells of the lead teams chiming in musical rhythm!
> Leo M. Weisenberger, Letter to the author.

THE PEOPLE of Tulare County have always loved trees. If any proof is needed, look at the thousands of acres of Kaweah delta farmland still dotted with specimens of the original oak woodland. It requires a considerable regard for sentiment or scenery to plow and irrigate around a thousand random oak trees for a hundred broiling summers.

They thought no less of the trees up there on the timber-thatched mountains. But just as Solomon needed timber for the building of the temple, so did they for their homes, schools, and churches. After all wasn't there plenty of timber for both houses and scenery? The forest was so big and the ax was so small.

Before we describe the individual sawmills, it will be useful to consider briefly how lumbering was done in those days.

<u>How It Was Done</u>. The period with which we are dealing was the one in which logging rapidly emerged from the strictly ax and bull-team method to the threshold of the power-logging of today. It is sometimes said that lumbering is essentially a "cut 'em down—cut 'em up" process. But in another sense, it

is primarily a transportation job. Efficiency, or the lack of it, in moving wood from stump to mill, and mill to customer, is what makes or breaks most operations.

In our area, getting the stuff down off a 6500-foot mountain to the 300-foot elevation of the San Joaquin Valley was the big problem. To reduce that problem as much as possible, the sawmills were taken to the woods. Elster says the lumber was dried for 30 days before it was hauled. In this way, only dry lumber had to be hauled down the steep crooked roads, leaving in the woods the bark, slabs, sawdust, and most of the sap. (Fifty per cent of the weight of green redwood lumber is water.) Even so, the cost of hauling lumber by horse and mule-team was so great that every alternative that human ingenuity could devise was attempted.

Mr. Dillon's wooden car-track (see Chapter VII and map) was the first such attempt in our area. Later, water-carrying flumes built of lumber were used to carry the green lumber (with the sawdust going along as a sort of lubricant and to get rid of it) from the mill as far toward market as possible. The first flume of which we have any knowledge was in use at least as early as 1883. It carried lumber about a half-mile around the high falls and down Rancheria Creek from the Rand-Haughton Mill to the road near the present Churchill cabin. Another carried Coburn's lumber from his mill down Bear Creek. It was in use from 1886 to about 1902. The Dillon car-track was replaced by a flume about 1900. All of the flumes of our area were flat-bottomed rather than "V" flumes, and carried lumber directly from the headsaw, one board at a time. Coburn's flume was 16 inches wide and 12 inches deep.

Lumber Hauling. Hauling lumber from the mills, or even from the "dumps" at the lower ends of the flumes, was no job for a greenhorn. Teams of as many as fourteen head of mules or horses were used. A photograph taken about 1905 shows A. J. "Jack" Doty hauling two wagons loaded high with lumber. The printed caption says, "This load of lumber contains 27,653 ft. or 55,306 ft. of resaw." This, incidentally is four times the amount of lumber, in the form of green logs, that the average present-day logging truck hauls out of these same mountains over incomparably better roads. The average load of lumber was, of course, much smaller; nearer 10,000 than 27,000; Griggs says 1000 ft. per span of horses.

The load mentioned above was hauled from Elster's Mill down the old Frasier grade, now abandoned as a public road. The names of his horses were considered important enough to be recorded. If the hills could repeat Jack Doty's voice as they echoed it 60 years ago, these are the names that would ring out across the canyons and valleys from old Mountain Home to Porterville: "Coaley! Mabel!" (wheelers), "Jip! Puss!" (pointers), "Pat! Don!" (sixes) "Bud! Bert!" (eights), "Babe! Joe" (tens), "Bob! Keeno!" (leaders). Mabel was especially famous throughout the district for her intelligence and skill.

As an aside, we are in danger these days of forgetting how important animals were in our grandparents' time. Take the case of old Joe Street's dog. Old Cuff, a black collie, was as smart as lots of men and more use around the house than some women. He could do all the things other intelligent dogs could do such as taking borrowed articles back to their owners and returning with baskets of fresh-baked gifts for his bachelor master, and finding lost horse-shoes for horses lamed by "casting a shoe."

But Cuff, so the story goes, had a special talent. He could take a pot holder in his mouth and take the coffee-pot off the stove. Of course nothing could take the place of the wife young Joe Street had failed to win, but Old Cuff tried. No wonder old Joe asked to be, and was, buried next to Old Cuff there in the Milo community. And incidentally, Milo was named for a dog. But that is another story.

The mountain teamster had to be tops in a class of men that held high status in that day. They guided their long teams as much by word of mouth as by their "jerk line." Monroe Griggs' book (1955) describes the intricate operation in detail. Men who teamed lumber from the Mountain Home and Dillon Mills included Griggs and Joshua Seamonds, who hauled lumber from the Frasier mill; Perry Osborn, father of Oliver and Clyde Osborn of Springville, who hauled from the old Rand-Haughton, Coburn, and Enterprise Mills; Tom Pedigo and Clyde Tyler (for Coburn); Delbert Dillon, Jack and Elmer Doty, Marion Anderson and Bill Thompson of Springville; Sandy Woods; and many, many others.

To a good teamster the feeding and comfort of his team came first. With this in mind, Mrs. Ola Hubbs of Visalia says that her brothers, Jack and Elmer Doty, "thought more of

LUMBER HAULING

their horses than of themselves." They usually drove teams of eight 1600-pound horses hauling two wagons, the driver riding one of the wheel horses, with a jerk-line to the leaders and a rope to the brakes. The round trip to Porterville required four days.

These skilled men and their educated draft animals did not come cheap, and when the lumbermen heard about new traction engines used on some of the big ranches, they had to try them out. The first to be tried was a "Farquhar" made in Pennsylvania about 1888 and rated at 10 horsepower (Tulare County Historical Society, 1957).

It was bought, in a definitely used condition, in 1900 by J. W. Kyle. He intended to use it for hauling lumber from the Enterprise Mill above Mountain Home, but in taking it up the Frasier grade for its first trip something broke, and it rolled backwards off the road down the very steep, brushy mountainside above the Haughton place and "literally smashed to pieces, leaving nothing worth picking up but the boiler." (Tulare County Historical Society, 1950.)

This loss, however, could not stop the wheels of progress. The June 8, 1900 Porterville Enterprise noted that a heavier "steam-wagon" christened the "Springville" started from either the Dillon dump or Springville the previous Tuesday with 38,000 feet of lumber and "went up the Daunt hill without stopping." Later that month the newspaper reported that it was making three trips a week to "Roth's Spur" (now Strathmore), and that Art Young thought it was a "cracker-jack." They had to surface some of the roads with redwood bark so it could get traction. This machine was a "Trilby" apparently purchased from the Holt Manufacturing Company (Tulare County Historical Society, 1950).

The use of this machine probably did not outlast the Youngs' operation which "folded" a year later. Another attempt was made, however, in 1909 or '10, according to Mr. Marion A. Grosse of Fresno. This tractor was put into use by one of the Nofzigers, who were operating the Dillon Mill at that time. Mr. Grosse helped widen and strengthen the bridges before the tractor was brought in, but it broke them down anyway and set fires "all over the country." So it, too, went back to the farm, and horses and mules came back to the lumber wagons for a few more years.

From Stump to Mill. That is how they got the lumber from the mills to town. But first they had to get the timber from stump to mill; and that, too, was a problem in transportation to be solved by Yankee ingenuity.

It was the job of the "faller"* to lay the tree where the logs from it could be most easily moved. He brought his razor-edged ax to the job each morning carefully wrapped in cloth to protect it from rust and nicks. Scaffolding was built around the largest redwoods, while springboards were used on the medium-sized ones. The springboard notches are still plainly visible on the redwood stumps. Axes were used exclusively at least until the 1870's (Holbrook, 1938). Even when the crosscut saw was available it was a long time before it was used for making the undercut in falling a tree. The fallers held out to the last against the two-man "Swede fiddle."

There is at least one redwood tree standing on the Enterprise logging area near Shake Camp where logging was started about 1898, in which there is an undercut about eight feet long and three feet deep, and both upper and lower cuts were made straight, clean, and smooth with axes. John Amick and his sons Eli and Jimmy, Earl McDonald, Pete Camp, and Henry Talley were well-known timber fallers. They were not called "choppers" as was common in the Coast Redwood country and in early New England (McCulloch, 1958).

The bucker or "cross-cutter" manufactured logs from the clear part of the tree. At first the ax and the wood-auger were the only tools available, so the very large trees were usually left standing or worked up into smaller products such as posts and shakes. Logs of all species and sizes had to be made to "ride" as easily as possible so men called "peelers" peeled the bark from them to reduce weight and friction with the ground. The heaviest ones were bevelled ("sniped") on the nose-end. The "swamper" was a low-paid helper around the woods, who did such work as clearing brush, building roads, and helping more skilled men in chute-building and team skidding.

---

*Most western woodsmen are aware that the verb "to fell" is in the dictionary, but they would no more think of using it than they would of going slickshod in the woods. Let easterners and flatlanders call the man who cuts down trees a "feller" if they wish, but the word-smith who is hammering out this volume will stick with the colloquial transitive verb "to fall."

CHUTES

The next job was to drag, or roll, or carry, or "chute" the logs to the mill. The most primitive method was merely to hook on enough yoke of oxen to drag the log by "bull-strength."

By this time (1886-87), however, the "skid-road" was in use, as can be determined from remnants still visible on Bogus Creek. The skid-road or "cross-skids" consisted of short logs set in the ground crosswise to the direction of haul and just far enough apart that oxen could step between them. The term was also rather loosely used to refer to any trail or road on which draft animals were used to move logs. The term "on the skids," meaning just one step from the final end, comes from this type of logging.

Chutes. Replacing the "skids" or "skid-road" came the logging chute (also called "the skids" by some loggers) which was made by joining large poles (we would call them logs) end to end, usually in pairs. They made a shallow trough for the sawlogs to "ride" in. In a two-pole chute, the poles were hewn on one side so that the chute had a wide "V" in cross-section. In 3-pole and 4-pole chutes the poles made their own "V". The poles were laid on, and fitted and pegged to, heavy crossmembers embedded in the ground or mounted on timber trestles. There were "running chutes" which were so steep that logs would slide on them by force of gravity, and ordinary chutes on which logs had to be pulled by teams straining along a tow-path alongside. Boys were hired as "skid- or chute-greasers" to grease them to make the logs move better. The grease has preserved remnants of the Elster, Conlee, and Enterprise chutes from around 1900 to the present.

Basically, a chute was built like a railroad except that the "ties" were farther apart and the "rails" were close together. The main chutes branched and rebranched like railroads and crossed stream-channels and skirted rock outcrops on trestlework. It took a pair of timber fallers, working steadily, to provide one chute-builder with material, according to Jack Stansfield of Lindsay, who built chutes for Elster and Dillon mills as well as in the Millwood country. All of the North Tule mills by 1895 used chutes to bring the logs to the mill, judging from sections that can still be found above Dillon, Conlee, Enterprise, and Elster Mill sites. No record has been found in our area of logs being hauled by the heavy log wagons used extensively in Fresno County before 1885 (Hurt, 1941) nor of the use of "big wheels."

Bulls. Motive power in the 80's and 90's was primarily oxen, or "bulls" as they were always called (even though steers were often used). William Byron "Barney" Vincent probably was the last of the bull teamsters in the logging woods of this area. Men who worked at the Elster Mill with him in 1904 remember well how the woods rang with his shouts and the names of his bulls.

Their names seem to be remembered better than those of the human loggers. Dan and Brandy were the "wheel yoke," Dime (Diamond) and Landy the second yoke, and Rock and Tony the lead yoke. The bulls were guided by "Gee" (right) and "Haw" (left) and encouraged with a bull-whip, to the butt of which was fastened a goad made from a sharpened nail.

A skillful teamster could get perfectly coordinated action by "gording" the "near ox" with the nail at the same time the "off ox" received the snap of the whip. Strong and pointed profanity also seemed to help. Holbrook (1938) says, "By all odds the most important man of a woods crew. . . . was the bullwhacker. . . . When he raised his voice. . . . the very bark of the smaller firs was said to have smoked a moment, then curled up and fallen to the ground."

After rolling or skidding the logs one by one into the chute, the bull teams pulled the "turn" (sometimes as many as twenty logs) to the mill by means of a log chain "trail-hooked" to the rear log. Some of the other bull-teamsters of the Mountain Home area were a "Kit" Carson (Enterprise Mill), and Charles Doty and Charles Elster (Coburn and Dillon mills). For a vivid eyewitness account of bull-team logging in northern Tulare County see Mrs. Lizzie McGee's story (Tulare County Historical Society, 1950). Horses or mules were sometimes used to pull the logs after the bulls or donkey engines had gotten them into the chutes.

Mules were used at the Enterprise Mill in 1898-99. About that time the woodburning steam donkey engine was brought in. A photograph taken near the Coburn Mill shows a Dolbeer donkey practically identical to the one shown in Andrews' book (1954) as the first steam donkey ever built for logging. The Coburn, Dillon, Enterprise and Elster mills all had steam donkeys, using them either to skid logs to the chutes or, at the

TIMBER AND LUMBER

mill, to pull "turns" of logs down the chutes. "Donkey-punchers," "spool-tenders," firewood cutters, and line-horse drivers made up the crews of the donkey engines.

Timber and Lumber. Of course, only the best trees were cut. No crooked, conky, knot-studded trees for them. They wanted their logs "straight as an arrow, sound as a dollar, and smooth as a schoolmarm's leg," to quote one logger. Those with experience in such things say that is very high quality.

Very little fir or cedar was cut. The sapwood was slabbed off the pine and redwood. Both Mr. Earl Wortman and Jack Stansfield state that redwood was the mainstay of the southern Sierra sawmills. They could make money on redwood but often lost money on pine. Lumber sold for about $10 per thousand at the mill. Irvy Elster says H. F. Brey in 1903 paid Charles Elster $11 per thousand at the mill or $20 delivered in Porterville, grade 3 and better, all species. However, mills were often in financial straits and had to sell for less. Mrs. Irene Phillips says that when her husband worked for Coburn he had to walk or "ride the flume" down to the Coburn dump on Sundays and stack up his week's wages to dry, because he was paid in lumber.

Fence posts sold for about 5 cents each, stacked at roadside in the woods. Wages were good (about twice as much as farm wages, according to Clem Simpson) when they could be collected. He states that fallers and chute-builders, the highest-paid workmen in the woods other than teamsters, were paid as much as $80 per month, others down to $50.

Working Hours. Stansfield says the Central California Redwood Company mills about 1904 paid $30 to $50 per month and board for an 11-hour day. There were no clock-watchers in the woods. He tells the story that when he signed up to work at one of the mills, the timekeeper asked him if he had a watch. When he proudly showed him his new dollar Ingersoll the timekeeper took it and hung it in a case with dozens of other watches, saying, "The whistle blows at six, twelve, one, and six. You won't have any use for a watch here."

It is interesting to note that almost half of the wholesale value of lumber went to the teamsters for hauling it from the woods to Porterville. (This was without help from any Teamsters' Union too.) This shows how much progress has been made in transportation in sixty years, compared to the other phases of lumber manufacture, because now the haul to Porterville would make up only about one-eighth of the production cost of the lumber.

<u>Sawmill Shanty-towns</u>. The sawmill was a busy community. It included, besides the mill itself and its lumber drying yard, the blacksmith shop, cookhouse, an office and commissary (or "wanigan") where records were kept and wool socks and hickory shirts were sold, a barn for the work stock, a bunkhouse for single men (nearly as comfortable as the barn), and any number of family shanties scattered around the nearby woods. The traditional all-male lumber camp did not exist in the North Tule woods. The cooks were usually women, and the "hashers," girls.

At Mountain Home there was always a Saturday-night dance. Stage-wagons, vegetable peddlers, local butchers and meat peddlers, horse-and-burro rental stations, and saloon keepers made their goods and services available, so there was little need for anyone to go to town. On Sundays there was fishing, sight-seeing by buggy or from horse or burro-back, berry-picking, and on occasion, hunting of deer, bear, mountain quail and grouse. Mountain Home, during its sawmill days, had Fourth-of-July celebrations, summer-term schools for the children, a croquet court, even the "Mountain Home Glee Club" shown in an old photograph equipped with cowbells, a musical (?) saw, and other noise makers.

The sawmills of that time were much like the small mills of today, although today's mills do not employ "screw-turners" and "ratchet-setters" (who clamped the logs on the carriages) nor "sawdust herders." Two circular saws were used in the larger mills, one above the other. It has been said that there were only two men in the area who could "hammer a saw" to refit it, Avon Coburn and Rube Prescott.

<u>Fred Green's Story</u>. To sum up this chapter let Fred Green of Corning, California, a grandson of Nathan Dillon, tell how it was done at the Dillon Mill. The following is from a letter dated December 14, 1962 to his brother Eugene of Visalia.

"Dear Gene. . . .

"I worked at the mill the summer of 1904. . . . We cut

FRED GREEN'S STORY

mostly redwood; the big logs we blasted into quarters. Bored a hole with auger and put in a heavy charge of powder. The powder man was a Swede and I was his helper for a while.

"There was a big bull donkey at the mill that pulled logs in with a heavy cable and a lighter cable (back line) that took a short cut and pulled the heavy cable back out. There was one, or, two maybe, smaller donkeys that yarded the cut logs into the chute. . . .

"The boarding house was run by the Wrights, a father and two sons. The old man was dish-washer and flunky and one son was cook and the other was waiter. This was the time the tough timber beasts were coming out from Michigan and Minnesota. This waiter wasn't a big man (160 or 165). He was quick as a cat and had a kick like a mile. . . . He soon convinced those eastern loggers he was boss in the dining room after he worked 3 or 4 of them over.

"Jack Stansfield was the Superintendent and the bull of the woods. He was a husky man then. He had a man named Wilson, a kind of a straw boss in the woods, his brother in law, I believe.

"Mill was run by a steam engine. Engineer was one of the Cramers that lived somewhere around Milo. I lived in a cabin with three other men, a cabin or two below where Uncle Alma and Aunt Mary lived which made it nice for me as I was always welcome at their house. Aunt Mary often had a piece of pie or a plate of cookies for me, which is another reason I was always fond of her. I was 18 years old in 1904 and we worked 11 hours per day from 6 to 6. If we were working 2 miles in the woods, we had to walk out and be there by 6 a.m. Also came home on our own time. Consequently an 18-year-old kid was always hungry.

"Will Putnam ran the commissary and kept books for the company. Doesn't seem possible now, but several times during the summer Uncle Alma and I would walk over the mountain to Middle Tule, fish down to Long Meadows, eat lunch, fish back up and then each carry 75 or 100 trout home. Another 14 or 15 hour day. And when I think of timber fallers and log buckers pulling a 16 or 18 ft. saw for an 11 hour day, they did raise men in those days. Some of the above guys wouldn't weigh more than 150 lbs. Barney Vincent with his bulls (oxen) leveled the ground so the big trees wouldn't break when they fell. Also he hauled up the provisions from the dump.

"An old long-whiskered Yankee, we called Manila John, did the finish work on the logging chutes. For an adz he used a tempered and sharpened mattock and when he was finished it looked nearly like a plane had gone over them.

"Don't believe I mentioned that Uncle Alma was scaling the logs for the Dillon heirs. As I remember, the Dillon Estate got $2 per thousand, seems ridiculous.

"The saws were two five- or six-foot circulars one above the other. I was off-bearer in the mill the last work I did. They had trouble keeping an off-bearer. When a 2-inch cut was made 4 or 5 feet wide and 14 or 16 ft. long it was really heavy and it had to be let down on the live rollers as easy as possible. The trick was to get a shoulder under an edge and come down with it, but taking all the weight possible with your legs and I had good legs and got extra money. Then it went to the edger and into the flume and was stacked at the Dump. My arm is about broken. Never knew before you were so damned inquisitive. . . ."

"Fred"

CHAPTER XI

THE SAWMILLS

> I often wonder what man will do with the mountains---
> that is, with their utilizable, destructable garments. . . .
> The Sierra crop of conifers is ripe and will no doubt be
> speedily harvested. . . .
> Hyde's Mill, booming and moaning like a bad ghost,
> has destroyed many a fine tree from this wood. . . .
> John Muir, *John of the Mountains*.

LUMBERING HAD ITS DAY in the Mountain Home area mainly from 1885 to 1905 and in Dillonwood somewhat longer. A heavy demand for lumber and good prices gave it a flying start (Barton, 1907). The sawmills for which timber was cut in the North Tule area during this period, in the order of their appearance, were the Dillon, Coburn, Frasier, Conlee, Enterprise, and Elster mills. Shake and shingle mills were operated by Howe, Cherbbonno, Dillon, and others.

The Dillon Mill on the North Fork of the Tule continued until 1900 in the Section 10 location to which it was moved in 1875 (see Chapter VII). It is uncertain as to how much it was operated after the attempt by James and Wagy in 1882. County records include a 10-year lease of land and sawmill by Dillon to James Morton in the fall of 1891. Nothing more is known of Mr. Morton.

The mill was taken over in 1900 by the Dillonwood Lumber Company of which J. W. Young, his sons Art and Ed, E. T. Cosper, and J. O. Hickman were the organizers. They moved the mill to its final location in Section 4 and replaced Alonzo Elster's old wooden car-track with a lumber flume. According to the Porterville Enterprise of August 17, 1900, "The new Dillonwood sawmill is a solid neat concern and the way it is put up indicates that the Young Brothers are in the sawing business there to remain awhile." On November 23, 1900, the Youngs reported 400,000 board feet of lumber cut that season.

In spite of a good start, however, the Youngs went broke in two years. The mill apparently was taken back by Mr. Dillon and sold to a Los Angeles sash and door manufacturer, a Mr. Hughes, for $5000. He operated it for one year (probably 1902) cutting sugar pine only and then "dumped it on Brey and Boole" (according to the late Clem Simpson of Springville). They rebuilt the mill to 40,000 feet per day capacity (Quinn, 1905) and employed forty to fifty men in the mill and thirty-five to fifty in woods and road work. H. F. Brey was a hard-working retail lumberman of Porterville. F. A. Boole was manager of the big redwood operation at Millwood. Apparently Mr. Brey acted as the manager of the Dillon Mill and they operated under the name of the Central California Redwood Company, a corporation capitalized at $200,000 on January 20, 1903 in Los Angeles, the other incorporators being Ed. W. Davies and Fred R. Harris of Los Angeles, and Wm. G. Uridge of Fresno. Mr. Uridge was apparently the most active at the woods end of the operation. Mr. Brey's daughter, Alberta, active now in the Brey-Wright Lumber Company of Porterville, says that Uridge was treasurer, and that the funds of the company later disappeared about the same time that he skipped out to Paris. Before this happened, however, the company operated the mill for a year or more (including the year 1904), and leased A. M. Coburn's box factory and planing mill in Springville (Menefee and Dodge, 1913). They also bought the Enterprise, Frasier, and Elster mills in the Mountain Home area. Machinery from the Enterprise mill was used to make major additions to the Dillon Mill. At the peak of operations their lumber hauling required sixteen ten-horse teams. The company "blew up" in 1906.

Marion Grosse of Fresno says Frank Boole was running the mill in 1908, and a Mr. Nofziger in 1909. F. U. and D. I. Nofziger, A. B. Tirrill, and V. L. Ward organized a second Tule River Lumber Company in Pasadena in 1909 with a capital stock of $50,000. F. U. Nofziger was president of the Porterville Northeastern Railroad which was completed to Springville in 1911 (Tulare County Chamber of Commerce, 1959). The lumber company was defunct by 1912 but the railroad continued to operate between Springville and Porterville until 1935.

A report by Harold Schutt (Tulare County Historical Society, 1950) says the Dillon mill operated "on and off until 1914." Oliver Watkins of Porterville says he hauled out the mill

THE COBURN MILL 85

machinery in the early 1930's and sold it for junk. Some of the lumber rollers are still in use at Earl Wortman's mill near Exeter.

It would be impossible to name all of the hundreds (possibly thousands) of men and women who worked at the Dillon mills. Several are named above and in Fred Green's letter in Chapter X. Jack Stansfield says he cruised the finest sugar pine timber there that he ever saw. He helped move the Enterprise and Frasier mills down the Frasier grade to Dillonwood in 1903 or 1904. Practically everyone who worked at the Mountain Home mills also worked at one time or another at the Dillon mill. Stansfield says Brey had a crew of about seventy-five men when he ran it. Ola Hubbs says the cookhouse fed over a hundred men at times. Old photographs show an amazing network of log chutes leading to the mill, and flumes carrying water to the various buildings and carrying the lumber away. There was a locally famous steam-powered "splitter" from the Enterprise Mill, which was really a rip saw set up to cut the large redwood logs lengthwise into sections small enough that they could be put on the log carriage.

Nathan Dillon and his son Alma ran a shingle mill below the sawmill. Marion Grosse graded lumber at the Dillon "dump" in 1909. The lumber was not piled as we see it done nowadays. It was set up X-fashion over a long fence-like rack. Dillon's youngest daughter, Mrs. Veda McCoy, "waited table" in the cookhouse about 1906 when Frank Boole was running the show. Mrs. Edna Meddick ran the cookhouse for Nofziger. John Spees of Springville rode the carriage in the mill at that time. Pete McKiernan of Porterville drove the "line horse" for the donkey engine about 1904, and he also recalls that as a boy in 1900 he rode a horse behind the "steam wagon" that hauled lumber from Springville to Strathmore. His job was to put out the fires set by that machine. Clem Simpson says he worked in the woods and cut 12-foot logs off the tops of the high redwood stumps left from the logging of still earlier days. Malvin Duncan of Porterville says his father, Charlie Duncan, drove the chute team for Young Brothers and previous owners of this mill, and that the Youngs paid their men with 30-day time slips that could be cashed only at a 10% discount.

<u>The Coburn Mill</u>. Avon M. Coburn was the principal local competitor of the various Dillon mills and operated more

continuously and more successfully than any other millman on the Tule. He was established in the business before 1885 (see Chapter VII), having operated at sites near Churchill's and Frank Knowles'. To review briefly the early history of the mill Coburn bought in 1883 or '84, Coburn bought Rand's interest in it when it was located at the head of Rancheria Creek in the northwestern corner of what is now the State Forest. Rand and Haughton had bought it from James Kincaid.

In 1885 Coburn built a road from the Knowles site northeast to a small meadow on a fork of Bear Creek near the southeast corner of Section 35. The location was on E. W. Haughton's new timber claim. It has come to be known as Coburn's upper mill site. He built a flat-bottomed lumber flume to carry the lumber as described in Chapter X. His first lumber "dump" was just east of the present Mountain Home Conservation Camp, and the comparatively level spot where this camp was built was his drying yard for one year. The road is still in existence without any change of location, and the millsite is marked by the charred pilings of the mill, the remains of his steam engine (made in Troy, N. Y.), and the boiler still mounted on its masonry foundation. Some insulators still remain on trees near this location reminding us that Coburn here operated what was possibly the first telephone line in the county.

The mill was still at this location in 1892 when an exhibition tree was cut nearby for the 1893 Chicago World's Fair, but it probably was moved that year to a site a half-mile west and below, on the main Bear Creek. Part of the dam and the ditch he built to supply water to his flume are still in existence, and his blacksmith shop location is marked by the blacksmith's treasure heap of broken pieces of hand-forged logging and mill equipment. The old road to this site is also in its original location and is occasionally used by those who don't mind taking a chance. This mill was sold to Charlie Elster in 1901 and operated at the same location by him until he moved it up the creek in 1903.

Coburn built a box factory and planing mill at what is now Springville in 1890. It was powered by a large overshot waterwheel, located near the soda springs. According to his nephew, Allan Hodge of Springville, he filled orders for a variety of finished articles in this mill, as he was of an inventive turn of mind and an effective salesman.

THE COBURN MILL

According to Menefee and Dodge (1913), Coburn was the "father" of the village of Springville, though William G. Daunt had operated a store south of the present village since the 1860's. In 1889 Coburn bought a tract formerly a part of John and Newton Crabtree's sheep-ranching set-up, and the following year set aside eighteen acres as a townsite, and "sold lots and lumber on easy terms to his employees." The same year he also bought the store from Daunt, who, incidentally, was his father-in-law and one of the financial backers of his enterprises.

Coburn's integration of enterprises was probably one reason for his comparative success in the lumber business. However, even he went bankrupt about 1906, reportedly because he spent too much time and money trying to market his invention of a two-cycle engine. He deeded everything to his father-in-law. The planing mill burned about 1907. After the failure of the Uridge enterprises about 1906, Coburn was able to get back his old sawmill at Mountain Home but he did not operate it until 1920. This was in cooperation with Charlie Elster at a location near the present Ed Bace ranch. Coburn was killed in an automobile accident in 1921.

Not much is known about Coburn's crew at the upper site. An old photograph shows Charles Doty driving a bull-team there in 1892. Elmer E. Doty worked there, he has reported, as a "swamper for Clyde Tyler." Others who worked there at least long enough to help with the World's Fair tree in 1892 included Rass James, Tom Hilyard, Bob White, Ed Shuey, Sam and Frank Talley, and John Calkins (Doty, 1947). At the lower site Mrs. Irene Phillips of Visalia says that she cooked for a "short crew" the fall of 1894, and that Justin Burgess was woods foreman and Lon Phillips "screw-turner" in the mill. Her sister's husband ("Aut" Hubbs) and Clem Simpson worked in the woods. Loua Cherbbonno worked there in 1894. The donkey engine photo mentioned in Chapter X shows Harry Amick apparently the "donkey puncher" or engineer, Fred Dumont spool-tender, Bill Millinghausen driving the line horse (which took the cable out to the logs in the woods), Carr Wilson, and George Johnson.

At the Springville plant Peter Ferry was foreman in the mill and lumber yard, according to Marion Grosse. Coburn's business manager was Warner I. Hodge, who married Avon's sister Eva. The story is told that Eva used to bring the

Coburn payroll the twenty miles from Porterville to Springville every two weeks in her buckboard. For protection she carried an Iver Johnson 5-shot "38" revolver with which she was reported to be a crack shot. She also carried a loosely covered can of red pepper. No attempt was ever made to hijack the payroll. It may never have leaked out that she carried it.

Altogether the Coburn mills operated on Bear Creek from 1885 or '86 to 1902 or '03, about seventeen years. This includes all three locations and the five years or so that Elster ran it as lessee and owner at the lower millsite. About 480 acres were cut over, 60 acres of which is now State Forest and the remainder Sequoia National Forest land. Logging was by bull teams. At the lower site, chutes still can be found. Bull teams put the logs in the chutes; the Dolbeer donkey engine was used to pull them down the chutes to the mill.

As for the volume of timber cut, probably fifteen million would be a reasonable estimate for the seventeen years. The season was short because of snow and bad roads in winter and insufficient water to flume the lumber, at least in some summers (Porterville Enterprise, August 24, 1889). Coburn's was not a redwood operation. There are very few redwood stumps on his cut-over land, and most of them appear to have been cut for posts or stakes. Coburn's "cutovers" are now,

"Covered all over with timber,
Like hair on the back of a dog."

They include some of the best young sugar pine timber in the southern Sierra and much of it is now of sawlog size.

The Frasier Mill. Not much is known about where L. B. Frasier came from and still less about where he went. He is probably the "man named Fraiser" mentioned by Hurt (1941) as having sawed lumber with a double-circular steam-driven mill on Pine Ridge in Fresno County from 1883 to '85, logging with ox-teams and log wagons. C. B. Doty (1947) describes the moving of Frasier's mill on July 5, 1885 with five 6-horse and one 4-horse teams. He lists the moving crew as Fred Howe, Benton Breeding, Bill Breeding, Tom Hilyard, Bill Boyd, Dave Rogers, Burk Smith, C. B. Doty, and a man from Hanford. The Doty family took a leading part in building the new road, later called the Frasier grade, from the Bear Creek road to the new millsite, and undertook to build and operate a

Battle Mountain looking toward Blue Ridge from the Balch Park Road. The brushy bottomland in front and to the right of the mountain is where the Indians fortified themselves in 1856.

Seven of the mysterious "Indian Bathtubs" on Indian Rock south of Jordan Peak. Indians ground seeds for food in smaller holes nearby.

Plate 2

Trapper Frank Knowles' apple orchard and cabin. Both date from 1885 or earlier. Moses Mountain in background.

Remains of one stone fort put up by Indians near Battle Mountain in 1856. Photo 1961 on the Harry Scruggs ranch.

Plate 3

James Kincaid and family at the Hollow Log at "Summer Home" (now Balch Park). Kincaid is holding the horse ridden by his wife. Courtesy Mrs. William Dye.

A band of sheep headed for the "shepherd's empire" in the Sierra.

John Hassack in his Porterville saloon. Hassack was a sheepman at one time and Hassack Meadow was named for him.

The Centennial Stump in 1918. Charlie Elster and a man from Prof. A. E. Douglass' party. Laboratory studies at the Tree Ring Institute dated the central ring of this stump as 1202 B.C. The tree's age when it was cut about 1877 was therefore approximately 3080 years.

Plate 5

An early-day logging crew posed in front of a large sugar pine undercut for falling. Note fallers' equipment: double-bit axes, wedges, and even a device for sighting the direction of fall. The two fallers are standing on springboards. Note size of chips, a matter of pride among fallers. The man on horseback might be the owner of the timber.

Plate 6

A falling ax weighed about five pounds and a medium-sized Sierra redwood like this might weigh 500 tons, but Henry Talley (L.) and Earl McDonald are making the undercut into chips so large a boy could hardly lift one. The original photo, owned by Joe McDonald of Springville, is captioned "Cutting a Sequoia at Mountain Home about 1904."

Plate 7

Coburn's planing mill and factory in Springville on the Tule River, 1904.

Two men riding out to the woods in the "pig" pulled by the haul-back line of the donkey engine. The enormous size and volume of timber in this "3-pole" chute was necessary for handling the great redwood logs at Dillonwood. Photo by courtesy of Mrs. Ada Swisher.

Plate 8

Barney Vincent, the last of the oxdrivers, and three yoke of the "bulls" he loved and cursed. At left is Grove Elster, his swamper and "bull cook." (He fed the bulls.) This skid road and chute can still be traced on the mountainside above old Mountain Home. Photo by Gertrude Oldham, 1903.

One of the few photos available that shows the construction details of the 2-pole chute. Barney Vincent's bullteam pulling logs down it, 1903.

Plate 9

The Frasier Mill—1885-87. The proud bull teamster with the fancy shirt among the "bulls" is Barney Vincent in his younger days. Note "sniped" logs on deck at right.

Plate 10

Moving big logs by donkey engine about 1904, probably at Dillonwood by **Central California Redwood Co.** In the upper photo they have rolled the log so that one end is on the chute. The donkey engine at left will now pull the other end so the log will ride in the chute. Note the two line horses in r. foreground. Lower photo: Rolling the log away from its original lay. Logs were peeled to make them slide more easily. Photos from "Scenes from Tulare County" by A. R. Moore, undated.

Plate 11

Probably the first steam donkey engine in the Tule River timber and one of the first models made. The line horse pulled the cable and chain dogs (hanging in foreground) into the woods and the engine pulled them in, the "spool tender" coiling the cable on the ground as the engine turned the "capstan" or spool. Peeled logs in background. From left, Bill Millinghausen, Fred Dumont, Harry Amick, George Johnson, and Carr Wilson. Coburn Mill before 1903. Courtesy Tul. County Hist. Soc.

A record load of lumber in Porterville about 1905. Jack Doty is shown riding a wheel horse of his famous 12-horse team hauling 27,653 board feet of Mountain Home lumber. Tul. County Hist. Soc.

Plate 12

Plate 13

The Dillon Mill in 1904. Note branched log chute leading from woods to mill, water flumes serving each building, and the lumber flume across foreground carrying lumber away from the mill building at right. Huge redwood logs are on the chute, lost off the chute at center left, and on mill rollway. Courtesy M. A. Grosse.

Building a sawmill. This is the Elster or Mountain Home mill. The workmen are placed as carefully as a lodge of master Masons. The owner, C. J. Elster, is setting an example of industriousness by driving a spike at extreme left, and the millwright, Rube Prescott, is in a supervising attitude at upper right. The skilled artisans occupy the center positions with the "badges of office": the blacksmith, "Bull" Bowie, holds one of the wood augers he has sharpened, and the boss carpenter, Jim McDonald, holds a carpenter's square. The rank and file are, from upper left: "Kelly" Blake, Pete McKiernan, Fred Prescott, Wally Rutherford, Cliff Breeding, Grove Elster, Lon Oldham, Bill Campbell. Mandy Bright, and Fred Dumont. George Johnson is in shadow at lower left (1903). Courtesy Tul. County Hist. Soc.

Plate 14

Frasier Mill. Built by L. B. Frasier in 1885 $\frac{1}{4}$ mile east of the Mountain Home resort.

Plate 15

Elster Mill in operation—1904. The old Mountain Home hotel is barely visible under trees left of center. Photo by Gertrude Oldham.

A hunting party leaving the Mountain Home resort about 1905, each with his own pack animal. From left: Art McFarland, Will Burford, Charlie Hardiman. Courtesy of Otis Brough family.

Plate 16

This fire-blackened stump is specially prized by scientists as the source of the oldest redwood material known, over 3000 years. In left and right backgrounds are the remains of chutes from the woods to the Enterprise Mill rollway across center of picture. Taken in 1918 by Professor A. E. Douglass, University of Arizona Tree Ring Institute.

Same scene as above photographed in 1962, 44 years later. Men standing on the same stump.

Plate 17

These fence posts—5000 of them—were cut from one redwood tree. Puzzle: How many men, women, and children can you find? Photo by Mrs. Gertrude Oldham about 1903 at Mountain Home.

Coburn's sawmill boiler at upper mill site abandoned in 1892. Author is shown with it as it appeared in 1958, overgrown with incense cedar trees.

Plate 18

Mountain Home as sketched for Thompson's Atlas (1892). Houses at r. are at Sunset Point. Mount Moses in distance at left; Maggie's "twin peaks" in far background.

Summer Home as sketched for Thompson's Atlas. Road at left later became the Balch Park Road. The pack animals are starting out on the old trail to Doyle's Soda Springs. Balch Park's Hollow Log and Lady Alice tree are in center foreground. Scene at upper left is now "Doyles" above Camp Wishon. Circled inset represents the Wishbone Tree above the State Forest headquarters.

Plate 19

Commercialism ran rampart among the Sierra redwoods in the early days. This photo was taken about 1892 to advertise Pear's Soap. The name of the girl holding the soap is unknown. Miss Rose Cherbbonno and her sisters Addie and Agnes are apparently illustrating the various uses of the soap. The tree is the old "Nero Tree" and the flume carries water to the Cherbbonno "shake mill."

Croquet game at the old Mountain Home resort while it was operated by the Dotys, about 1890.

The Mountain Home Stage at Mountain Home resort probably about 1890. Courtesy of Bill Rodgers.

Plate 21

The photographer's legend on this photo is, "Eye-witnesses to the preparation and departure of the World's Fair Big Tree being photographed upon the stump at Mammoth Forest. Tabor Photo, San Francisco." The only persons identified are Mrs. Martha McFarland (back-right) with her two sons, Art and Ed, Van Dorman (the entrepreneur of the operation, front-R.), and probably Ben Harper (longest beard). The upright slab is half of the floor piece of the exhibit.

Same stump as above, 1958. Near Coburn's upper millsite. Photo by Harold Schutt.

Nero Tree. Everybody and his brother had their picture taken at the Nero Tree before it was cut down for a proposed exhibit at the Saint Louis Exposition of 1904. From left, lower row seated: Burr Breeding, Oren McKiearnan, Mrs. John Kirk, Mrs. Burr Breeding and child Beryl, Don Elson (boy with hat in hand), and Everett Kirk. Next row, standing: Andy Baker, Jim McDonald, Fred Talley, Mandy Bright, Irvy Elster (on burro), John Kirk (holding Viona, later Mrs. Nicholson of Exeter), John M. McKiearnan and wife, Grandma Rebecca Elster, Cora or Eva Baker, Bertha Kirk (Mrs. Sterrett), Velma Breeding (Mrs. Manier of Porterville), Georgia Elson(?), and Earl McDonald. Seated on scaffolding: Fred Dumont, Grove Elster, Bud Futrell, William Campbell, Anita Kirk (Mrs. Hamilton), Orval Moore, Harold Moore of Lindsay, Fern Moore, Earl Vincent, Mrs. Bob Moore, Della Kirk (Mrs. Carder of Visalia), and Lora Elster(?). Standing on the scaffolding: Clifford Breeding, Pete McKiearnan, Frank "Kelly" Blake, Bill Millinghausen, Sam Duckwall (or Kibbler), Eula McDonald (Mrs. Roy Smith), Ellen Cross (or Ethel Elster), Lola Akin (Mrs. Hairs), Wallace Rutherford, Chester Doyle (above Wallace), and Bob Moore (below Chester).

Plate 23

Nero Stump—1955. Located in Frasier Mill Public Campground, Mountain Home State Forest. Courtesy Tul. County Hist. Soc.

A long crosscut saw made about 1903 by Disston Saw Co. for John M. McKiearnan for some unknown purpose.

Fourth-of-July parade complete with horse-drawn float. Mountain Home resort 1904. Photo by Mrs. Oldham.

Plate 24

Forest Rangers and other U. S. Forest Service personnel from the Sequoia and Kern national forests, 1913. Left to right, back row: Dr. E. P. Meinecke (pathologist), Al Redstone (Hume Lake Ranger), Regional Forester Coeurt DuBois, Bill Rushing (Supervisor, Kern), Purdy, Evans, Parkinson. Second row: Ranger Bob Beard, unknown, Riggles(?), Freeborn(?), Sedman Wynne (Ass't. Supervisor, Sequoia), A. B. Patterson (Supervisor, Sequoia), two unknown men. Front row: Ranger Bill Derby of Springville district, veteran ranger Jess Brown, Ranger Bill Klingan, Ranger Irving Wooford, unknown, Lloyd Allen, F. P. Cunningham (Ass't. Supervisor, Kern), an unknown man. Courtesy Mrs. Della Cunningham.

Spike buck and gray squirrel, Mountain Home State Forest.

Plate 25

A big one at Dillonwood, 1951. The white ring is the sapwood just inside the thick fire-resistant bark. Spread-eagled on the end of the log is Pat McDonald, owner of the logging company.

Rolling a redwood cross-section from a windfall for the Los Angeles County Fair. Mountain Home State Forest, 1961.

Plate 26

Timber fallers using modern chainsaw and old-time springboards to fall a giant Sierra redwood. Dillonwood. P. H. McDonald Logging Co.

Tractor logging in mixed timber. Immature redwoods in center. Mountain Home State Forest, 1961. Photo by Robert Johnson.

Plate 27

Hedrick's Mill about 1943 at Hedrick Pond, Mountain Home State Forest. Courtesy of Gordon Tate.

Same grove of redwoods as is shown back of the mill in the above photo, 1958. The automobile is on the Balch Park Road.

Plate 28

A redwood downed with dynamite about 1944 on Michigan Trust Company land. Note man between down tree and standing tree. U. S. Forest Service Photo #436259.

Breakage in the brittle Sierra redwood. About 1944, Michigan Trust Company. U. S. Forest Service Photo #436236.

Plate 29

Deputy State Forester Cecil E. Metcalf on the steps of "Hercules," the room tree at Camp Lena. Courtesy of Fresno Bee.

Forestry Camp Superintendent Ray Little (r.) and the author on a cross-section of a windfall redwood that was cut in 1961 for a permanent exhibit at the Los Angeles State Fairgrounds, Pomona. Photo by Tulare County Chamber of Commerce.

Plate 30

"There are four rules for living in the mountains; let there be no formation in trees, no arrangement in rocks, no sumptuousness in the living house, and no contrivance in the human heart"
    One Hundred Proverbs, Translated from the Chinese by Lin Yutang.

Plate 31

THE FRASIER MILL

hotel that they named "Mountain Home" a quarter of a mile from the new mill. All this and the building of the new mill took place in 1885. Both mill and hotel were located on the route of the old Dennison or Coso Trail.

Frasier built a good-sized mill that summer on a 27-acre piece of ground acquired from E. W. Haughton in the northeast "40" of Section 35. It has been rated at 40,000 feet per day capacity. J. P. Ford was the millwright. All that remains now to mark the site is the masonry foundation for the boiler and a piece of the smokestack.

Apparently Frasier bit off more than he could chew and soon got into deep financial difficulties. The county records give some clues to the sequence of events. Frasier bought a 160-acre timber tract in Section 26 from Mary Seamonds. The purchases of both millsite and timber were recorded July 7, 1885, the same month he moved his mill. The next April he mortgaged some of his property to Allen McFadgen for $6000 and some to other persons. On September 16, 1886 John M. McKiearnan leased the mill from Frasier for $240 per month. In the spring of 1887 there were several suits against Frasier and on October 29 of that year foreclosure proceedings were filed by Wm. J. Newport of Hanford, at that time a County Supervisor. W. H. H. Hart and T. J. Hilliard (Tom Hilyard?) were also involved. Title was recorded in the name of Jarrard, Newport, and E. R. Pease on October 4, 1888. The Tulare County Historical Society (1950) says the mill burned in 1888. Something of value must have remained, however, because the Porterville Enterprise of May 18, 1889 says Newport informed the editor "that he has disposed of his Frasier Mill to Moore and Smith of Stockton." Moore and Smith were two of the incorporators of the Tule River Lumber Company later that year, and were promoters of lumbering in northern Tulare County. The property was recorded in the name of Smith Comstock on December 13, 1889, and later transferred to the Tule River Lumber Company.

Frasier's troubles with his road stirred up a little excitement. This road, now a private road still usable by jeeps, leaves the Bear Creek road near the corner common to Sections 4, 5, 8, and 9, and climbs northeasterly 2800 feet in about 3 miles to the Balch Park Road. The latter road follows the old Frasier Road the remaining one and a half miles to the millsite. As stated above, Frasier was supposed to pay for the construction of this road. Men of the Doty family were prominent among the construction crew, and apparently were not fully paid for their work. When Frasier came back in 1889 to attempt to establish his road as a toll road, the move was bitterly contested by the Dotys (for whom it was the access road to their Mountain Home hotel and resort) and others. Frasier went to law. He lost his suit but built a gate across the road anyway (Porterville Enterprise July 20, 1889), and "backed up the same by shotgun law." The county then proceeded to open it and Frasier to close it again. Whereupon Frasier was arrested for obstructing a public road, and, "On Sunday the 21st, a mob of about 10 men armed with axes and rifles proceeded from Mountain Home to Camp Two on the Frasier road and proceeded to demolish the gate and fences. Mr. Frasier on the following Tuesday went to Visalia and swore out a warrant for the arrest of the three principals of the gang, Clyde Tyler, W. Sullivan, and A. J. Doty, Jr. . . . . They . . . . were dismissed on the grounds that the road was a public highway." (Porterville Enterprise, August 3, 1889) It appears from later items in the paper that Frasier did not give up until finally the county got out of the controversy by appointing "viewers," who recommended that the road be abandoned as a county road. However it was in use for tourist traffic and millions of board feet of lumber for thirty years after "the Frasier war."

    Few of the men who worked at the Frasier mill are known, other than those already mentioned. A photo of the mill shows "Barney" Vincent and six or seven yoke of oxen prominently displayed. Trying to find correct names and dates connected with this mill is like looking for one of last winter's icicles.

    In spite of all its troubles, the Frasier Mill must have put out quite a lot of lumber. It was rated at 40,000 board feet per day, or twice the size of Coburn's largest mill. About 190 acres of timber on some of the best land in the area were cut over, all of which is now State Forest land. It is estimated that three million per year were cut in 1886 and '87 and enough in 1885 and '88 to make a total of 7,500,000 feet. Judging from the stumps on the area, much of the cut was redwood. Only the medium-sized redwoods were cut for sawlogs. Many large trees still stand.

    Logging was by bull-teams dragging logs down cross-skid skidroads, some of which can still be found. The best

CONLEE MILL

second-growth redwood stands on the State Forest are on the Frasier cut-overs. Their acreage probably exceeds 80 acres and on some areas the sawtimber volume already exceeds 40,000 board feet per acre.

Conlee Mill. The best known site of the Frank Conlee Mill is at the lower end of Brownie Meadow on the line between Sections 34 and 35. It was operating in 1893, according to a picture owned by Mrs. Irene Dillon and captioned "Conlee Mill Crew--1893." About twenty-five men are in the picture, including several of the Talley men, and A. J. Doty, Sr. and his son Jack. This mill operated also in Section 27 where Williams, Walker, and Hooker have built summer cabins. The last setting for this mill was just east of the Mountain Home Guard Station beside the station's spring in Section 35.

At some stage in the history of the Conlee Mill it was taken over by Bill Breeding and his wife's brother, Joe Washburn, for what was a common reason for such transfers of ownership, namely, the inability of Conlee to pay his help what he owed them. Others who worked at this mill in 1893 included Lon Phillips, Charlie Elster, Milt Hubbs, and a man named Nelson. One of the reasons this mill is remembered is that an epidemic of "mountain fever" (probably typhoid) hit the crew in the fall of 1893, and Nelson died of it.

Conlee's Brownie Meadow millsite has been but little disturbed since it was abandoned. The stone foundation for the boiler and the cedar pilings that supported the mill machinery are still in fair condition. A bearing that was found at the site has this inscription cast on the grease box lid: "EF. LIGHT'S PAT. AUG. 21, 1886." All of the Conlee cut-over land is now in federal or private ownership. No evidence as to how Conlee did his logging at the first two sites has been found, but above the Guard Station site there is a two-pole chute in good condition.

Enterprise Mill. With the organization of the Enterprise Lumber Company, sawmilling in our area entered a new phase. The previous mills operated as personally-owned or partnership projects and were financed locally. The Enterprise was financed by "outsiders," four of the five incorporators giving San Francisco as their address, viz., Allen McFadgen, Wm. F. Beeson, Matilda Watson Burns, and Hazle Burns. The other was Albert Brown of Fresno. The company was incorporated June 4, 1898 with a capital stock of only $20,000, but it "let a lot of daylight into the swamp" during its two or three years of operation. A Tulare County Historical Society article (1950) says the mill was built in 1897. It could have been built by McFadgen or others before incorporation of the company.

Apparently the most active member of the company was Allen McFadgen. He had been taking mortgages on and buying choice timberland in this upper Bear Creek country since 1886. The Enterprise Lumber Company, however, owned only the so-called "McFadgen 80" in Section 25 along the Camp Lena Road where a large sawdust pile now marks the mill site. After this land was logged, operations were extended toward the north where G. E. Guerne owned 120 acres. Guerne is reported to have furnished much of the financing for the Enterprise operation. He purchased his land from J. J. Doyle. The Doyle and McFadgen families were connected by the marriage of Allen's son Jack to Doyle's daughter Ruby.

The Enterprise Mill, rated at thirty to forty thousand board feet per day, was larger than any other Mountain Home mill except possibly the Frasier Mill. It removed all of the timber from the "McFadgen 80" and possibly 40 acres additional. The total cut was in the neighborhood of ten million feet, which would be over 80,000 feet per acre. This would be possible only because it was redwood country. Hauling the lumber from the mill to railroad employed fourteen teamsters.

Logging was by 2-pole chutes with bull-teams at first. Later donkey engines skidded the logs up-hill to the chutes, and mule teams moved them along the chutes. Men reported to have worked in the woods included John Amick, woods boss; Eli and Jimmy Amick, fallers; Clem Simpson, chute-builder; Harry Amick, spool-tender on the donkey engine; and "Kit" Carson, bull-teamster. Charlie Duncan helped build the mill and the road to it (the present Camp Lena road). J. W. Kyle was the sawyer in the mill, Lee Lindsay the book-keeper, and Otis Brough and Chet Ainsworth worked there. Both Dan McFadgen and Loua Cherbbonno are reported to have run the shingle-mill. Elmer and Jack Doty hauled lumber to Porterville. Frank Knowles' last days in the woods were spent as watchman at the mill about 1900 (Tulare County Historical Society, 1955).

The logging chutes can still be traced from the millsite to the far corners of the logged areas. Woods operations must have closed down "between two suns," because at the upper end of one of the chutes, redwood logs are still lying in all stages of manufacture, skidding, and chuting; and trees are standing that are undercut for falling. Even springboards used by the fallers have been found cached in a burned-out "goose-pen" in a redwood tree.

All of the logged area is now a part of the Mountain Home State Forest. The "McFadgen 80" is covered with a somewhat patchy stand of Shasta red fir up to three feet in diameter (the only stand of this species on the Forest), and white fir and redwood. Very few of the original redwoods were left standing. The Enterprise Lumber Company got into financial difficulties and ceased operations in 1901, according to Elster. The corporation ceased to exist in 1904.

The history of the Enterprise follows a familiar pattern of high hopes, followed by a few years of big operations getting out the logs and teaming huge quantities of lumber down the mountain. Then the longer skidding distances to the mill increased the costs, or a slump in lumber prices occurred, or both, and the operation "folded." It was lots of fun while it lasted, but nobody got rich. Lots of men had man-sized jobs (and usually got paid), and those who put up the money were prevented from spending it foolishly.

Or were they? Sometimes when we look at the "McFadgen 80" to-day (even though it is well-forested and very inspiring in its way); imagine it in all its primeval majesty at the time it became soneone's "timber claim;" recall how it passed from owner to owner like a baseball; and see, at last, how the tall red giants, lopped and dismembered, rode down the greased chutes and through the screaming saws to become just another eight million units of merchandise; we wonder if, as a people, we didn't "pay too much for our whistle."

<u>Elster or Mountain Home Mill.</u> With the exception of the Dillon Mill, the last mill to cut lumber in the North Tule area until many years later was the one built by Charles Elster just a few hundred yards above the old Mountain Home Hotel. Many people are still alive who worked on this operation or had close contact with it.

Although Elster took over Coburn's Mill at its lower site in 1901 or earlier, his operations there have been included as part of the Coburn Mill history. The following account includes only the operations of the rebuilt mill at the Mountain Home site. Elster moved the mill by donkey engine a distance of about one and a quarter miles up Bear Creek without benefit of any road, Carr Wilson being boss of the moving crew. This was in 1903. A good photograph of the new mill under construction indicates that it was a well-built structure. Rube Prescott was the millwright and James McDonald, head carpenter.

The mill began operations in late 1903 or in 1904, the lumber going to H. F. Brey in Porterville. By 1904 Elster was in serious financial difficulties but fortunately was able to sell out to the Central California Redwood Company represented by Wm. G. Uridge, an Englishman of Fresno who was engaged in promoting several enterprises ranging from Fresno city subdivisions to brick plants. According to a rumor, supported by County records, he was financed by the wealthy Captain Leslie J. Holman of England. Archibald Kains was also involved. Uridge's reputation as a lumberman was that "he couldn't even make good sawdust," but he paid off promptly in real cash money and therefore had no difficulty in hiring help. Jack Stansfield was in charge of the woods operations for Uridge. Jack says that woods and mill together used thirty to forty men, but that the payroll was loaded with considerable extra help because Uridge was short on lumbering experience but long on cash. The "honeymoon" did not last long. After about three months the mill was closed and the crew moved to the company's Dillonwood operation.

Some additional people who worked at this mill included J. W. Kyle, a good sawyer who moved there from the Enterprise Mill; Lock Cramer, edgerman; Harry Amick, Aut Hubbs, Milt Hubbs, and Malvin Duncan, donkey crew; Clem Simpson and Elmer and Jack Doty, lumber teamsters; "Barney" Vincent, bull-teamster in the woods; and, last but by no means least, Mrs. Minnie Elster, Mrs. Ola Hubbs, and Mrs. Edna Meddick, cooks.

The site of this mill is now a pleasant meadow in the turn of the Balch Park Road just above old Mountain Home. Some of the logging chutes are still in recognizable condition. Nearly all of the seventy acres of logged area is now a part of the

SPLIT PRODUCTS 95

State Forest. The days of the Elster Mill were lively ones because of the large numbers of summer campers and tourists who joined with the woods workers and their families in a variety of activities ranging from sight-seeing in the redwoods to square-dancing.

Split Products. If all the redwood cut for fence material, shingles, shakes, and stakes in the North Tule area were added, the total volume would probably exceed that cut for lumber. Before the introduction of the cross-cut saw (which differs from all other saws in having "raker-teeth" interspersed between the cutting teeth) the only tool that could make a straight cross-cut in a large log, without wasting an inordinate amount in chips, was the wood auger. The first redwood felled in the Sierra was severed from the stump by boring closely parallel holes with an auger. This was in Calaveras County. Since this method of crosscutting was a lot of work, the largest redwood trees prior to about 1885 were worked up into fence posts, rails, fence palings, grape stakes, split boards, roof shakes, and shingle bolts. Large sugar pine trees were made into split roof shakes also, usually in preference to redwood. For splitting redwood it early became common practice to bore a line of two-inch auger-holes down the top of a long section of a large prostrate tree, fill them with black powder, and then split the section lengthwise by setting off simultaneous charges in all of the holes.

Shake and Shingle Mills. The first mill to saw roof material in this area was apparently the Dillon shingle mill shown on the township plat of 1883 on the North Fork above "Dillon's." The second was probably the one operated by Warren Howe (about 1886-90) on the site of the Frasier Mill. An ingenious operation was begun about 1892 in what is now the southern part of the Frasier Mill Campground. The owner and operator was a French-Canadian, Loua Laurence Cherbbonno (pronounced Shav'eno by Tulare County people). Cherbbonno's first mill was a "shake mill," according to his son, Loua "Ben" Cherbbonno of Lake Hughes, California, who supplied the following information about his family's activities. The shakes were not tapered as shingles are, and they were uniformly six inches wide and three feet long with two grooves cut in each to carry the run-off. The statement was made that Cherbbonno invented the first power drag saw. A photograph of it shows that it was powered by a stationary steam engine from which a long shaft was run parallel to the felled tree. The sawworks was moved along this shaft three feet at a time as each cut was completed.

Loua Cherbbonno's wife, Louisa, in addition to mothering her family of six children, was the engineer for the family's mills, and on one occasion, for the railroad locomotive from Porterville to Bakersfield.

<u>Summary</u>. The exploitive lumbering of this period--before the word "forestry" was ever heard in the forest--had a tendency to saw off the limb it sat on. Its balance was precarious at best, and the year 1905 saw the end of lumbering in the Mountain Home area until thirty-five years later. The seven sawmills of this area plus the several Dillon Mills had cut most of the trees growing on about 2200 acres, only about one-thirtieth of our whole four-township area. The acreage was divided about equally between Mountain Home and Dillonwood. The operators had logged, sawed, and hauled to town an estimated forty million board feet from the former area and fifty million from the latter. In addition, millions of feet of redwood that had borne the bulky snows of unnumbered Sierra winters found their way to valley ranches to be festooned with barbed wire or grape vines.

CHAPTER XII

MOUNTAIN RETREATS FOR THE SAN JOAQUIN

And this our life, exempt from public haunt,
Finds tongues in trees, books in running brooks,
Sermons in stones, and good in everything.
                    Shakespeare, *As You Like It.*

WHENEVER A FEW MEN built a road to reach timber and started to cut it down, hundreds followed that road up the mountain to find relaxation under the "tall uncut," and to watch the axes flash and the bull-teams weave and strain. The hot, poorly-drained, fever-ridden San Joaquin Valley of that day made the mountains attractive to almost everyone. Many families not otherwise financially able to go to the mountains combined the rough, dusty work of logging or post-cutting with an all-summer family camping trip. So it was that lumbering opened up the mountains for recreation. The end of each lumber road tended to become a center for tourists, and a jumping-off-place for sheepmen and prospectors. Therefore, it was logical that someone should put up a hotel and summer resort at such spots if other conditions were favorable. Thus wood and wonder became twin commercial assets of the Mammoth Forest country.

Among the popular places to go in the Sierras to escape the burning heat of the San Joaquin Valley were the resorts of Mountain Home and Summer Home. Before their time, in the 1870's, Dillon Mill was a camping center and a place to start from for trips to Mineral King and Mt. Whitney. Summer Home was located in what is now Balch Park; Mountain Home a mile north. There were other resorts, but it is believed that these were the most popular of any south of Yosemite in the late 1880's and '90's. Apparently Mountain Home was favored by Visalia people and Summer Home by folks from Tulare.

To serve as a "half-way house" for travellers to these resorts (because you could not drive from valley towns to the mountains in one day) John Gaffney and his sister Ann opened a hotel on the Bear Creek Road at Rancheria.

<u>Mountain Home</u>. Three children may as well have the honor of being the pioneers of the Mountain Home settlement. It was this way, according to Mrs. Ola Hubbs and her sister Mrs. Irene Phillips, both of Visalia. One spring day in 1886 the Doty family, headed by Andrew Jackson Doty and wife Sarah with their horses and wagons were slowly climbing the newly built Frasier Road toward a timber claim they were to call Mountain Home. They all knew every foot of the road. The previous summer they, with others, had built it for L. B. Frasier, successively moving their residence from Camp 1 (above Rancheria) to Camp 2 (Iron Gulch), Camp 3 (now the J. E. Bace apple ranch), Camp 4 (later the Vincent place, but now reverted to dense woods), and Frasier's mill. With some of the first lumber Frasier sawed after he built his mill at the end of the road, Doty had put up that fall a story-and-a-half cabin a quarter-mile down the road. Now, on this day the following spring, as all the older members of the family labored to repair the winter's damage to the road and get the wagons over it, three of the children, Jack aged 11, Irene 9, and Ola 7, too young to help much but too old to be babied, walked ahead as youngsters will. While troubles delayed the main caravan, anticipation speeded the feet of the young vanguard and before they knew it they had arrived at the cabin. It was too dark to go back and the rest of the family were unable to reach the cabin that night, so young Jack took charge and quieted the fears of his sisters. In this way they started, as far as any record to the contrary goes, the first family residence in the Bear Creek redwood area.

So much for the beginnings. In 1886 Frasier's Mill was the center of things in the mountains and the Mountain Home Hotel was the place to stay. One of the first things father Doty did was to build a croquet court. Dances were held every Saturday night, the usual program through the night being two square dances, a waltz, then two more squares and a polka. Alvin Slocum and his son George played for the dances on their home-made violins made of redwood and manzanita wood. What the oldsters seem to remember best about Mountain Home in those days is not the big trees or the big goings-on in the woods, but the big good times they had there as youngsters.

MOUNTAIN HOME

Sarah Doty set a table that attracted both tourists and workingmen. The Doty children acted as guides to show visitors the big trees, the Centennial and "California" stumps, the "Indian bathtubs," and other nearby attractions. An old friend of the Dotys, Jesse Hoskins, also acted as a guide and an active friend of the Mammoth Forest wonderland. Hoskins, a well-to-do rancher of the Lindsay area, acquired the present Camp Lena 80 acres even before Frasier and Doty got their timber tracts. He spent his summers there, his best remembered activity being the carving of the room in the "Hercules" tree, a project he started in 1897.

Hoskins had a hobby of naming the redwood trees and fastening name-plates on them about twenty feet above the ground. Instead of naming trees for generals and presidents, as was done further north, his selections were more classical. Methuselah, Hercules, the Adam Tree, Napoleon Bonaparte (a hollow log), and Nero were probably named by him, as well as certain others named for less famous people such as Anne and Anthony, the Seven Sisters, Uncle Lon, Old Clumpy Zeb, etc. Mrs. Keagle (1946) has described his activities in detail.

Another guide who showed people the sights in the period around 1904 was Mrs. Gertrude Oldham, now of Springville. She was an excellent amateur photographer and would take pictures for the visitors, and develop and print them for her customers immediately in her cabin at Elster's Mill. Her "trade mark" in most of the pictures is her two small sons, because, she says, she could always keep her eye on them that way.

Schools were held for the children in summer. Mrs. Keagle says that the top of the Centennial Stump was used for a classroom. Other reports are that classes were held in a room in Doty's hotel. (There are also reports that the Centennial Stump was used for a dance floor, and that four squares at a time were danced in the Hollow log at Balch Park. It is quite possible that such things did take place once in a while as a stunt, but the peaked shape of the Centennial Stump is such that dancing on it would be close kin to mountain climbing.) In March 1890, according to the <u>Porterville Enterprise</u>, Doty was trying to get a post office named "Doty" at his hotel.

It has been estimated by people who frequented the area during these years that 1000 to 2500 people could sometimes be found camping in the Mountain Home—Summer Home—Camp

Lena area during the heat of the summer. Much of the San Joaquin Valley was devoted to grain farming in those days, and after harvest was over in June, the ranchers brought their families to the mountains. Often they occupied their time by making redwood fence posts, grape stakes, and roof shakes or working for the lumber mills. At first most campers just "slept out" but after a terrific rainstorm in August 1887 soaked several hundred of them there was a sudden increase in the demand for the local lumber for building rough cabins. The report is that 700 people were camped at Mountain Home that August plus several hundred at Summer Home; and that 3 babies were born there that year (Tulare County Historical Society, 1950).

Joe Doctor, in his book Shotguns on Sunday, quotes one of Chester Doyle's stories about a tent saloon operated by the notorious Porterville outlaw, Jim McKinney. It was just up the road from the Mountain Home Hotel, but the Doty family want it known that it definitely had no cooperation from A. J. and Sarah Doty. The story is that McKinney sold his whole outfit one day to some of his customers for $150. Then they all proceeded to celebrate the deal with free drinks on the house and soon "liquidated" the business. In 1889 G. W. Thompson opened a saloon east of the Old Frasier Mill. This was probably to capitalize on the thirst of the Enterprise Mill sawdust savages.

A regular stagecoach line served both Mountain Home and Summer Home for at least one summer. The photograph we have of it does not bear much resemblance to the stagecoaches of the television westerns. It apparently carried both freight and passengers.

Now, to repeat the old sad story about Mountain Home as a place to get rich. The Doty's had to mortgage their 160-acre timber claim, hotel, store, and cabins to make ends meet and in 1894 lost the property to L. J. Redfield, a hotel man of Porterville, who operated it for a while. During the years of the Elster mill, which operated adjacent to the resort from 1903 to 1906, it was apparently tied in to some extent with Elster's and Uridge's operations. Mr. and Mrs. Ben T. Sickles ran the hotel and store for a year or two. The buildings were used for a variety of purposes until they were torn down about 1947. George and Irene Dillon and Charlie Spangler were two

A BIOGRAPHICAL NOTE

of the later operators of the store, finally closing it down in 1941.

A Biographical Note about the Doty family is not out of place here. A. J. and Sarah came to Hanford in 1882 from Oregon by way of Lake County, California. One of their twelve children died of malaria their first summer in that mosquito-ridden area near the great swamps around Tulare Lake. This no doubt influenced the family to look for a location in the mountains. The older boy, Charles, had preceded the family to California and had worked for Frasier at one of his Fresno County mills. So the family "threw in" with Frasier on his road building job and located their summer resort near his new mill, as has been mentioned. Among the boys in the family, Charles married Rosa Burgund; Elmer, Myrtle Manier; Jack, Hope Crabtree; and Moses, Clara Frame. Daughter Philindia married Fred Wells; Irene, Lon Phillips; Ola, "Aut" Hubbs; and Clara, J. Curry. Third and fourth generations are, needless to say, numerous and no doubt a credit to their ancestors.

Summer Home. John J. Doyle's old summer resort now makes up the main part of Tulare County's Balch Park. Doyle acquired a part of it very early, probably 1885. It was then known as Talbot Meadow and the creek running through it was called "Old Indian Bathtub Creek," according to the 1884 Official Tulare County Map. Doyle called it "Summer Home."

Doyle, as a man of twenty-seven, came to Tulare County in 1871 from Indiana. He was involved in the "Mussel Slough Tragedy" north of Hanford, and served eight months in prison. Because public sentiment in this affair was all on his side, he was treated more as a paying guest than as a prisoner. He had been the leader among the settlers in their land dispute with the extremely unpopular Southern Pacific railroad (Smith, 1939). He married Lillie Alice Holser. The almost perfectly proportioned redwood tree called Lady Alice near the Hollow Log was named for her.

Doyle was an enterprising man. He was one of the first to take up land in the Mountain Home area. A nephew, Wilbur Doyle, homesteaded the small prairie or "canada," at Doyle Springs just above the present Wishon Resort. For a few years after 1888 or '89 John Doyle made his home there, moving to Porterville in 1892 or '93. Summer Home, however, was his hope as a summer resort and real estate development. At one time he is reported to have sold 125 lots there. Once, during a drouth year, he started a ditch to bring water from the Wishon Fork just below Redwood Crossing to supplement the springs at Summer Home. The few hundred feet he dug are still discernable there.

The Hollow Log was the center attraction at Summer Home and everybody had to have his picture taken in front of it. Clinton T. Brown, the early-day sheepman, is the first visitor there of whom we have any record. His name and the date, 1870, were carved on the log (Tulare County Chamber of Commerce, 1959). Of course the place was probably a favorite camping spot for Indians, and for prospectors, sheepmen, and others who travelled Jordan's Trail before Brown came along. Doyle built a road to it from near Knowles' cabin about 1886, locating it so it ran through the "Wishbone Tree." He traveled by pack trail between Summer Home and Doyle Springs, where he planted an apple orchard. At first he packed his crop by horses or mules up this trail and, it is said, used the Hollow Log for an apple cellar. Chester Doyle, his only son, used to tell about how they lived at one time in two rooms inside this log. Other attractions on the property include the "Cosper Stump," a hollow redwood stump about 30 feet high which was reportedly used once as a two-story dwelling; and the "Sawed-Off-Tree," a large dead redwood that is supposed to have been sawed completely through, but is still standing on the stump in seeming defiance of the law of gravity and all the winds that blow.

<u>Exhibition Trees</u>. About every fourteen years, apparently, an urge would build up in John M. McKiearnan and he would have to go up to Mountain Home and cut another redwood tree for exhibition. McKiearnan, according to his only surviving son, Pete McKiernan (Pete has simplified the spelling of the family name) of Porterville, was a Tule River rancher who "took a flyer" once in a while in timber and mining ventures. He was one of the three partners who financed the Centennial Tree venture about 1877, described in Chapter VIII. In 1889 and again in 1903 he made other unsuccessful attempts to get redwood trees to national or international expositions.

THE CALIFORNIA TREE

<u>The California Tree</u>. The 1889 project is partially covered by "From Files of the <u>Porterville Enterprise</u>" (Tulare County Historical Society, 1950). According to the August 3, 1889 issue of this newspaper, John McKiearnan "felled the large tree at Mountain Home that he is to take to Europe, last Friday evening just at dusk. Many were disappointed, as only two persons saw it fall. Many campers had gone from Summer Home to watch the work for the past weeks." The October 5 issue states:

"At last the big tree 'California', which Messers McKiearnan and Davidson have been cutting down in the Redwood forest above Frazier, is ready and will be underway to visit those places where anything from the "wild west" will be welcomed as a curiosity, and shortly, those narrow-minded sceptics who have never seen a genuine Giant of the Redwood groves will have to admit that the fabulous stories told of the world famous but little seen Sequoia gigantea are true.

" 'California' was cut from a tree growing in the Redwood grove three quarters of a mile to the east of Frazier's mill and is a portion of a giant which grew to a height of some 300 feet and measured some 76 feet in circumference at the base. It has been cut into eight separate pieces each weighing some 1,200 to 1,500 pounds.

"John McKiearnan and three assistants accompanied "California" to Porterville, where they arrived at about 2:00 P.M. Tuesday. Monday morning will witness the departure of the stump for Visalia where it will be placed on exhibition for the forthcoming fair, after which it is bound for Tulare, Fresno, Merced, San Francisco, Sacramento, San Diego and San Bernardino. Eventually, it will bid a long farewell to its native state and will start for New Orleans via Texas."

That seems to be all we know about the "California" tree, except that its oddly-shaped stump still stands on State Forest land in Section 36 one-fourth mile south of the Centennial Stump. A rumor has it that McKiearnan gave up the project and sold the exhibit in Porterville where it met some totally unromantic end.

<u>The Nero Tree</u>. In 1903 he tried again. This time he selected a very unusual specimen of a redwood tree. Across Bear Creek from the site of the Frasier Mill stood a large tree that was almost perfect on the outside but completely rotten on

the inside. A very small "crawl-hole" enabled a person to get inside the cavity created by the heart-rot. This tree, perhaps because of its totally degraded heart inside a regal exterior, had long been known as the "Nero Tree." The other exhibition trees had been prepared by laboriously hollowing out a tall stump from the top and then cutting "barrel staves" from the shell thus produced. The Nero Tree had the big advantage of being already hollow. Mrs. Gertrude Oldham says that Mc-Kiearnan planned to send the shell of the stump to the 1904 International Exposition in St. Louis, Missouri. He went as far as to clean out the stump and started to cut off the shell from the inside, as may be seen now by anyone who wishes to crawl inside or climb over its twenty-foot-high rim. But the story is that John got an answer about then from St. Louis telling him what space would cost at the exposition and he decided he couldn't swing it. It is probably the only tree-stump in existence that looks as though an attempt had been made to chop the tree down from the inside. And in a sense that was what was attempted.

The Van Doorman or World's Fair Big Tree. The batting average of McKiearnan and his partners in getting redwood trees actually exhibited outside of California appears to be exactly zero. One other tree was cut that may have gotten further, but for this tree, too, we can find no record of where it actually went. This was the so-called "World's Fair Big Tree" of the Coburn Mill vicinity. (Not the World's Fair Tree from the "Chicago Stump" in the General Grant area that actually reached the "Columbian Exposition" or Chicago World's Fair.)

The Porterville Enterprise of March 11, 1892 carried an item stating that, "J. J. Doyle, in company with three other gentlemen from Tulare, was up in the redwoods above Coburn's Mill last Thursday and Friday. . . . to obtain one for the World's Fair. They have not made any selection yet." Some time later a man named Neal Van Doorman (or Dorman) came to superintend the cutting, hauling, and shipping of the exhibition parts of this tree. The operation is well documented by photographs by Tabor of San Francisco, all labeled "Mammoth Forest." Charles Doty (1947) says that Andy Baker, a local man, had charge of the job and that he (Doty) hauled the body of the tree to Coburn's Mill. A widely distributed bull-team photo on a calendar attests to the latter. Los Tulares carries

THE VAN DOORMAN TREE

an article (Tulare County Historical Society, 1950) about Van Doorman and his tree; and in <u>Pen Pictures of the Garden of the World</u> (Lewis, 1892) there is quite a detailed description of the Neal Van Doorman exhibit which is stated to have already reached San Francisco on its way to Chicago. We quote a paragraph:

"The entire piece of wood consists of sixteen sections as follows: The lower section is one foot in height by twenty feet in diameter, all in one solid cut, weighing 19,725 pounds. This will be arranged as a floor, placed on nine elegantly carved and enormous pedestals made of the same tree. The next is seven feet in height by twenty feet in diameter, which is hollowed out and will be placed on the floor cut. The whole of this remarkable curiosity will form a sort of hall and will accommodate about 100 people and will be entered by a swinging door made out of one of the portions of the second section."

CHAPTER XIII

MOVES TOWARD FOREST CONSERVATION

    Earth's crammed with heaven, and every common
shrub afire with God,
      But only he who sees takes off his shoes.
                             Elizabeth Barrett Browning.

    LOGGING OF THE REDWOODS, from the very first, was considered by many people to be a crime and a sacrilege. Gradually these people gained converts and the weight of their sentiments for preservation of the remaining groves began to count in the political balances. Rensch and Hoover (1933) state that the movement began in 1878 and was led by the Tulare County newspaper editor, historian, and conservationist, Col. George W. Stewart of Visalia. Opposing him and his group were certain lumber interests who were trying to block up their holdings in the Sierra redwood belt. We see now that the fight for preservation was lost on some battlefields and won on others. In the Converse Basin country in Tulare County capitalists Smith and Moore were there "firstest with the mostest" and they and their successors logged nearly every redwood except the Boole tree. In some of the best groves in the Tule River watershed the fight was lost to Smith Comstock and the Tule River Lumber Company, but relatively little of the private land there was logged. To the everlasting credit of Visalia's Colonel Stewart, victories were won and great areas saved in the Sequoia National Park, Grant's Grove, and what is now the Sequoia and Sierra National Forests.
    An interesting but little-noted account of the origins of the conservation movement is described beginning on page 273 of Small's (1926) <u>History of Tulare County</u>. Colonel Stewart began as early as 1878 (the year Congress passed the notoriously abused Timber and Stone Act) to print numerous references in his <u>Visalia Weekly Delta</u> to the need of conserving Sierra

Nevada forests, especially what was then called the Fresno-Tulare grove of redwoods. Small's history says that General John F. Miller, U.S. Senator from California, was made acquainted with the problem, and in 1881 he "introduced in the U.S. Senate a bill providing for the creation of a national park and reservation, covering the territory from the Middle Fork of Kings River to the North Fork of Tule River (thereby including all or part of our North Tule forest area) and extending eastward to the summit of the Sierra Nevada Range, but it failed to pass."

In 1880, due to efforts of J. D. Hyde, register, and Tipton Lindsay, receiver, of the U.S. Land Office in Visalia, Visalia people succeeded in getting the U.S. Surveyor-General for California (who had been a Visalian also) to suspend from entry the land in or near what later became the General Grant National Park. Again in 1885, according to Small, Tulare County people prevented this area from being sold under the Timber and Stone Act. If any special attempt was ever made to keep the North Tule redwoods in public ownership, it did not succeed, because, as we have seen, that area was taken by storm by Timber and Stone Act entrymen in 1884 and '85.

Small goes on to say that in 1885 "and each year thereafter the Delta made urgent appeals for the preservation of the forest." The issue of September 12, 1889 "viewed with alarm" as follows:

"The Federal Government should take measures at once to preserve the forest and prevent the Sierras from becoming a range of bare, verdureless, stony peaks like some portions of the Coast Mountains, that were at one time heavily forested, and thus prevent the western slope of the range and the broad plain below from being subject to successive floods and droughts."

The account further describes how Courtney Talbot (mentioned previously as an early-day land-owner in the Mountain Home country and a county supervisor) "upon reading this editorial. . . . brought the matter to the attention of the Tulare Grange, and a resolution was adopted providing for a committee of Grange members and others to meet in Visalia on the 9th of October to decide upon some plan for inducing Congress to take action in the matter."

To make a long story a little shorter, after various meetings were held at Visalia and Fresno and numerous committees formed, the job came back to Colonel Stewart as committee secretary (and to Tipton Lindsay as chairman) to prepare a memorial to Congress in which was to be designated the lands that were to be requested for the new park or reservation. The original idea of the committee was to include only a tract "in the vicinity of Mount Whitney," but the daring and imagination of Stewart and Lindsay shows forth in the map they finally prepared. It embraced practically all of the forested country from Yosemite southward into Kern County. The area, four million acres, was essentially that now contained in the Sierra and Sequoia National Forests and the Sequoia and Kings Canyon National Parks. The account is concluded by Small, with pardonable pride and a considerable amount of truth, as follows:

"There was at that time no provisions for the creation of forest reserves by Congress, by the President, or otherwise, and while no immediate action was taken on the petition. . . ., it has been said that, owing to the request of a large and definitely described area, it caused the preparation, presentation, and passage of the act of March 3, 1891, authorizing the President of the U.S. to set apart and reserve public lands wholly or in part covered with timber and undergrowth as public reserves."

It took four more years of battle, however, before all of the petitioned area was reserved. The Secretary of the Interior, because of pressures from certain private interests, had a habit of suddenly releasing for entry selected choice lands that had previously been suspended. In some cases the Visalia people by telegrams and political contacts were able to stop these releases, but finally they decided the cat-and-mouse game could only be stopped by making a determined drive for a national park in part of the area. This drive succeeded when, on September 25, 1890, an act was passed that set apart the "Sequoia National Reserve" (later designated the Sequoia National Park).

National Forests. It was more than two years after establishment of the Sequoia park before the Sequoia National Forest had its inception. On February 14, 1893, President Benjamin Harrison proclaimed as a forest reserve "the area described in the petition prepared in Visalia in 1889, the first request of its kind praying for the reservation of a specific area of forest-covered lands" (Small, 1926). Excluded, of

THE TIDE TURNS

course, were the new national parks, the lands that had slipped into private ownership, and some other exclusions. It was the second forest reserve in California and the thirteenth in the U.S. (Clar, 1959). Included in this reserve were the public lands in our North Tule area. It was named the "Sierra Forest Reserve" and C. S. Newhall of the U.S. General Land Office was the first supervisor with headquarters in Fresno. He had a few rangers to assist him. Troops were sent from Sequoia National Park to post notices warning sheepmen against trespassing. In the years around 1899-1900, Harrison White, brother of early-day Mountain Home sheepman Huffman White, was supervisor. In 1905 an agency named the "Forest Service" was set up in the U.S. Department of Agriculture, and two years later the "forest reserves" were re-designated as "national forests."

<u>The Tide Turns</u>. For 40 years the tide of private landownership surged in successive waves over the Tulare County foothills, into the timbered coves, and over the first redwood-studded ridges and plateaus. By the turn of the century, however, it was beating itself out against the rocky crags of hard economic facts. The steep, high, rugged mountains made lumbering expensive and the railroads brought stiff competition from more favored forest lands. Public sentiment, too, had its effect and placed some hurdles in the way of exploitation. Before 1900 the tide of private ownership had reached its peak and was even beginning to fall back. In the Mountain Home and Dillonwood areas, almost all of the private land that had not been purchased by the Tule River Lumber Company or by Dillon or Canty was acquired by certain traders (including Senator E. O. Miller and James A. Hannah) and transferred to what was then the Sierra Forest Reserve in exchange for "scrip" negotiable for purchase of public lands elsewhere. This was made in accordance with the "forest lieu clause" of a Federal Act of June 4, 1897. The transfers back to public ownership took place mostly in 1899, and included lands that had been logged by the old Hubbs and Wetherbee water power mill, Rand and Haughton, Coburn, and Conlee, as well as some areas that had not been cut over.

Along with the decline of the lumbering industry there was a decline in the resort business on private lands. The question is often asked, why did people stop going to Mountain Home?

There is no one simple answer, but the answers stem from the fact that, at least at Mountain Home, the recreation business on private land was dependent on the lumber business. Lumber built the roads, lumber maintained them, and lumber and fence posts furnished the income for many of the resort customers. When lumber could not make a profit, the summer resorts failed too. Giant Forest began to compete with Mountain Home for recreationists after 1903, because of the completion of a road to Sequoia National Park that year. The coming of the automobile had something to do with it, too. The old Frasier road was a tough one for a "Model T." The Mountain Home area did not revive as an important recreational area until the opening of the one-way "control road" to Balch Park about 1929.

PART THREE

THE PUBLIC ASSUMES RESPONSIBILITY

> Woods have tongues
>   As walls have ears.
>
>   Tennyson, *Idylls of the King.*

IF WE WERE to characterize the three time-periods of this story in terms of sounds, we might describe Part One as beginning with the crack of Frank Knowles' rifle sighted on a fat doe, or with the roar of Captain Livingston's howitzer at Battle Mountain. The timber faller's axe, the sawmill "booming and moaning like a bad ghost," and the eternal baa-aa-ing of sheep, would open the second period and continue through it. But now as we begin the third period, relative quiet prevails throughout the Mammoth Forest country. As the noise of sawmills and sheep diminish, the old primeval sounds are heard more clearly again: waterfalls and running streams; the calls, wing-whispers, and footfalls of the untamed; and wind, thunder, and fire. And on summer nights happy human voices still sing old songs around a dozen circled campfires.

But before the close of this final period, the prevailing quiet was broken by the loudest uproar of all; blasts that shook more people than those within earshot. But we are getting ahead of our story.

The first two parts of this book told the story of the North Tule people and their rugged mountains during the half-century between the mid-eighteen-fifties and 1905, with some "flashbacks" to earlier years. Now we propose to relate the highlights as we see them for the forty succeeding years, plus a few items of later date. We shall do it by grouping the stories under three headings relating to the Sequoia National Forest, Balch Park, and the private and state-owned land.

CHAPTER XIV

UNCLE SAM'S FORESTERS

> In the administration of the forest reserves, it must be clearly borne in mind that all the land is to be devoted to its most productive use for the permanent good of the whole people. . . . You will see to it that the water, wood, and forage of the reserves are conserved and wisely used for the benefit of the home-builders first of all . . . and where conflicting interests must be reconciled, the question will always be decided from the standpoint of the greatest good of the greatest number in the long run.
>
> Secretary of Agriculture James Wilson.

ON FEBRUARY 1, 1905, the responsibility for management of the nation's still new "forest reserves" was transferred from the U.S. Department of the Interior to the Department of Agriculture. The letter from which Secretary Wilson's quotation was taken was written to Gifford Pinchot, who assumed on that day the direction of the new "Forest Service."

At that time, the privately-owned lands of our "Mammoth Forest," were intermingled with lands of the Sierra Forest Reserve, a four-million-acre mountain empire. Charles H. Shinn was the Head Ranger with headquarters at North Fork, Fresno County. Before the year 1905 was out, young William B. Greeley, who later was to be the third Chief Forester of the United States, had ridden the length of the Sierra, checking on timber sales in progress.

The next year, to quote Mr. Greeley (1951), "I was made supervisor of the Sierra South, a forest of some $2\frac{1}{2}$ million acres between Kings River and Tehachapi Pass. . . . Headquarters was at a ranger station on a canyon flat well up within the forest, to which we brought our water by the bucket from a spring across the foot bridge. There I lived with my two saddle horses, my pack mule, and big sheep dog. The World was mine!"

SHEEP VS. CATTLE

Greeley tells a story on himself in which an old cattleman once taught him a lesson in democracy. At a gathering of cattlemen, he let himself get a little carried away with this "world is mine" stuff. One of those present casually made this observation: "When the young supervisor just now talked about his national forest, it sort of reminded me of the time when the old Devil took Jesus Christ to the top of a high mountain. He offered Christ all the kingdoms of the earth if he would fall down and worship Satan. All of 'em, mind you. The old s.o.b. didn't own a damn acre!" (Morgan, 1961).

Sheep vs. Cattle. How the Sequoia blazed the trail for the other national forests in getting control of sheep-grazing is graphically told by Greeley:

"Something had to be done about the trespassing sheep. They were eating out ranges allotted to permittees and putting the whole system of controlled grazing in disrepute. The rangers came in for a council of war and the campaign was planned. The migrating flocks left their winter pastures in the San Joaquin Valley in early spring, moved slowly through Walker Pass, and turned northward along the east slope of the Sierra. The herders. . . . played hide-and-seek with the forest rangers all summer.

"During the following winter, ... "No Trespass" signs in English and Spanish were put up every half mile. In the spring, secret lookouts were posted at points commanding the woolly line of march. There were enthusiastic volunteers for this duty from young cow punchers. Soon dust clouds carried the news that sheep were moving up Walker Pass, and headquarters had daily reports on the location of the lead flock. The district ranger ostentatiously left his station with his pack string, for a long trip over the high country.

"We waited for days while the sheep grazed along but never across the boundary. Then suddenly they moved in. We held back until the flocks had made a day's drive and a night's bedding within the national forest. Then the deputy supervisor rode up with three or four deputized marshals and a Basque interpreter. There were nine thousand sheep in trespass under care of a dozen herders and camp tenders. Through the bedlam of South European expostulations, the deputy made the boss herder pick three men to take care of the sheep. He arrested the others and marshaled them before the United States commissioner at Bakersfield.

"Suspicions of long standing were confirmed when top-flight attorneys from San Francisco magically appeared as counsel for the sheepherders. They represented some of the largest land companies in the state. They challenged the power of the Secretary of Agriculture and all his minions to control grazing on public lands of the United States. However, the commissioner bound the Basques over to the federal court and unwonted peace descended upon the south Sierra ranges. Many months later the Supreme Court received the case of United States vs. Grimaud, Cazazous, and Inda and settled for all time the authority of the Secretary to regulate grazing on national forests."

One day when Greeley was riding the trail with his chief, Mr. Pinchot renamed the Sierra South the "Sequoia National Forest" (Dana, 1958). The name was made official on July 2, 1908, by presidential proclamation.

June Eleven Claims. One of the first "headaches" handed the foresters on the new national forests was the Forest Homestead Act passed by Congress on June 11, 1906. This law permitted citizens to take up homesteads of 160 acres or less within the new national forests where the land was chiefly suitable for agriculture, and was not needed for public purposes.

The rangers immediately surveyed selected meadows and other areas that might be needed for guard and ranger stations. These were called administrative sites, and no less than 192 were reserved on the Sequoia.

There was a rush to file on the remaining lands especially by those who had already "squatted" on, or were using, lands within the Forest boundaries. On the Sequoia, 1103 "June 11 claims" were entered on the books from 1906 to 1916. In the Bear Creek drainage the former Bill Berry property (part of it became the O'Neill apple orchard) was taken up by J. P. "Pete" Planchon in 1914. Jake Abernathy acquired the present Luther Carl place the same way and Clayton E. Northrop at least a part of the Bear Creek Ranch. Most of the claims were taken care of by excluding from the Forest some marginal strips of land that were heavily "plastered" with these claims.

Greeley served as supervisor until June 1, 1908. He was followed by C. E. Sherman. In 1910, A. B. Patterson took over; in 1915, P. G. Redington; in 1916, Sedman W. Wynne; and in 1918, Frank P. Cunningham. Cunningham had left a cattle ranch to join the Forest Service and had come to the Kern in 1913. He ran the Sequoia for eighteen years. At that time, the Forest extended nearly to Owens Lake. Riding horseback from California Hot Springs across the whole Sierra to meet with a ranger at Lone Pine was commonplace then, according to Della Cunningham, his widow. For forest officers, "No trail too steep; no day too long," to quote a later supervisor, Norman Norris of Springville.

After the consolidation of the Sequoia and Kern National Forests in 1915, the summer headquarters of the Sequoia was at California Hot Springs until 1927 or '28. They did not always move in winter but when they did, they moved to Bakersfield (until 1920), then to Porterville.

Harry Wilkinson, a son-in-law of Jim McDonald, was an early-day ranger for the Springville area. From 1910 to 1919 Bill Derby, a Stanford University man, filled the post. His winter headquarters were at Rancheria, the Bear Creek site near the present "Sci-Con" camp. Wesley W. Snider, a native of Deer Creek and another of McDonald's sons-in-law, was next, serving from 1919 to 1936 and from 1943-46. Temporary winter headquarters in Springville were used from 1919 until the government station was built and occupied in 1932. In summer the headquarters was at Hossack Meadow, several miles by pack trail from the end of the road.

Supervisors following Cunningham were Joseph E. Elliott, 1935; Norman L. Norris, 1941; Elliott again in 1945; Paul W. Stathem, 1946; Jack McNutt, 1952; and Eldon E. Ball, 1953. Springville rangers after Snider to 1960 were Ray Stevenson, Paul Struble, Robert Cron, and Kenneth Fox.

The Mountain Home area was assigned a fire guard from about 1925 to 1945. Guards were hired for the fire season only, and were required to furnish their own horses. The first guard camp was at Crystal Spring with Alna Harding as fire guard. When Crystal Spring suddenly failed, the camp was moved to the new cabin at Balch Park, and later to the old buildings at Mountain Home. In 1934 the present Guard Station was built. Owen Rutherford was guard at the latter two locations.

<u>Grazing in the Mountain Home Area.</u> The sheep were gone from the North Tule forest country before 1917, according to a report prepared by Cunningham in that year. The national forest permitted 428 head of cattle, however, on their "Mountain Home Range" plus 35 head of settler's cattle and horses. He judged that the range would handle a hundred more. In addition Charles Gill ran 250 head on the Hume property in that area, paying $1.66 per head for the privilege, and 6,317 acres of other private land were estimated as carrying 195 head. This total of 908 cattle and horses is far in excess of its carrying capacity today and may well have been excessive then.

A stock driveway to Peck's Canyon went up the Frasier road to old Mountain Home and the Enterprise millsite, then up the Wishon Fork to Long Meadow and through Devil's Gap (a devil of a trail between Sheep Mountain and Maggie Mountain) to Peck's. Only cattle were permitted. The Dillon area also was given over to cattle, according to this report.

Norris says that the cattlemen and sheepmen fought it out for the range in the Wishon Fork and Peck's Canyon country in the drought year of 1898. The sheep cleaned the meadows and left no feed for the cattlemen's saddle-horses. Ranchers Ray Kincaid and Marion Anderson are said to have gotten their start in cattle that year by looking after the cattle of Duckwall and Kibbler for a share of the calf crop, and coming out well in spite of the tough competition with the sheepmen.

<u>Fires.</u> The earliest fire in the North Tule for which we have eyewitnesses was one that probably started from a moonshiner's still above what is now the Ed Bace ranch. That was in 1924, a bad fire year. It ran a mile northeast. Ranger Snider set up two camps of firefighters to control it.

The Lumreau and Dennison fires were in 1926, according to Norris. Lumreau was a lightning fire that burned the Bear Creek watershed from Lumreau Mountain to the Slick Rocks, 4,400 acres. In 1936, a fire on the Middle Fork side of that same ridge burned almost to Methuselah and Dogwood meadows but was controlled before it got into the redwoods. Another bad fire year was 1942.

The more recent fires in the Mountain Home area have been small. "Dude" Sutch of Springville says he had a few small fires from blasting redwood with black powder in the 1929-45 period. One burned 3600 of his posts. One year a bad lightning

FIRES

storm fired the dead tops of four big redwood trees at separate locations in the Mountain Home area. Snider, with Sutch's help, blasted down three of them to get at the fires so they could be put out.

CHAPTER XV

BALCH PARK

> This indenture . . .between A. C. Balch and Janet Jacks Balch husband and wife, Donors and the County of Tulare . . . , Donee, witnesseth: That . . . This conveyance is upon the conditions, restrictions, and limitations hereinafter set out . . . that the premises hereby conveyed shall forever be and remain in their present state and condition so far as may be possible and, subject to this controlling purpose, shall be made available for the use, pleasure and enjoyment of the general public.
>
> Deed conveying title of Balch Park to Tulare County, December 10, 1923.

MR. AND MRS. A. C. BALCH deserve full credit as the godparents of Balch Park, but John J. Doyle was its natural father. It was because of his ownership beginning in 1885 and his hopes for it as a summer home colony and resort that it escaped the fate of the "McFadgen 80" during the lumbering era. Doyle's story is told in Chapter XII.

In 1906 Doyle sold to the Mount Whitney Power and Light Company (later to be absorbed into the San Joaquin Light and Power Company) both his Doyle Springs homestead on the Wishon Fork and his beloved "Summer Home" property, now Balch Park. The power company had plans to build a flume from a power site on the Nelson Fork of the Tule and planned to cut the Summer Home redwoods for flume lumber. However, Mrs. John Hays Hammond, a large stockholder in the company, was taken to see the property and she was so impressed by the 200 giant redwood trees on it that she prevailed on the company not to cut or sell them for lumber. (An additional factor in saving the trees was that the company lost out to the Edison Company on water rights and abandoned their flume project.) Mr. Doyle also was anxious to see the redwoods saved and when he heard that Mr. and Mrs. Balch were interested in buying park properties to donate to the public, he was instrumental in

BALCH PARK ROAD

getting them to buy his old Summer Home from the power company for the purpose of donating it to the County as a park. Then he had to go to the county officials and get them to agree to accept and administer it. The sale to Mr. and Mrs. Balch was made in 1923. In 1927-29 the County completed their part of the bargain by building the road that is now called the Balch Park Road. On December 23, 1930 the County Supervisors officially accepted the property and named it Balch Park.

Allan C. Balch, incidentally, was an engineer who had graduated from Cornell University in 1889. By 1912 he was a vice-president of the San Joaquin Light and Power Company at Los Angeles, and he continued to be active in the business until his retirement.

Balch Park Road. The new road to Balch Park opened a route that had been little used since Charles Wilson's oxteams lugged the first sawmill up a long ridge into the "South Pinery" at Happy Camp about 1870. It was a somewhat roundabout way and unverified tales are told of political intrigue and personal antagonisms that may explain why the shorter, drier, possibly less expensive route up Bear Creek, or a route up from Camp Wishon, was not chosen.

Mr. Mack Bland, of Fresno, was in charge of the construction work. He and his partner, Arch Stewart, had contracted with Tulare County to build the road for about $50,000. The road was laid out by the late County Engineer Lawrence Moye. ("A prince of a man," Bland says.) Mr. and Mrs. Frank Negus ran the cookhouse. Bill Osborn was graderman. The job was started in March 1927 and was not finished until 1929. Grades were limited to 8% by the Balch deed, and curves to 40-foot radius by the County. Mr. Bland says that in building the road they measured grades on old lumber roads as steep as 37%.

Troubles included a camp-robbing mountain lion, half-wild pigs, and an old mountaineer whom we shall call George. This character tried to stop the road from going across his claim, and was accused of collecting dynamite and tools from the road job. He had the reputation of being able to make away with other peoples' property in very ingenious ways. According to a story retold by Mr. Bland, George once laid a trail of barley from his place to that of Jake Abernathy, a neighbor, so his (George's) hogs would go over to Abernathy's and fatten on his barley while Jake was away on a camping trip in the back country. The nearest that anyone ever came to catching him was when Charley Elster caught the heel of his shoe in a bear trap. Elster had set the trap in his orchard to catch the thief who had been stealing his apples. Don Witt says that George's neighbors credited him with sitting up on his hill ranch and noting when any of them left for town. Then he would go down and let their cattle out so he could pick them up as illegal loose stock under the No-Fence Law.

George was commonly accused of getting his meat supply by shooting the range hogs and calves of his lowland neighbors, in accord with the time-honored tradition of predatory mountaineers:

> "The mountain sheep are sweeter,
> But the valley sheep are fatter;
> And so we deemed it meeter
> To carry off the latter."

But it was wild game that got him into trouble with the law. He shot a mountain sheep, an illegal act in California. His general good luck prevailed, however. The judge threw the case out of court because the warden had charged George with shooting a male mountain sheep, whereas the University zoologists to whom the horns and fleece were sent for identification, pronounced the animal a female.

George even figured out how to run his old truck without buying gasoline. The trick, according to Wes Snider, is to run out of gas in a narrow place where nobody can get by without giving you some gasoline.

The new Balch Park Road was all narrow. For at least twelve years after its completion it was operated as a "control road." Uphill traffic was permitted only at four-hour intervals starting at six a.m., and downhill traffic at the same intervals starting at eight a.m. Anyone "breaking control" was subject to a court-imposed fine. "Control stations" were manned at both ends of the section that included the thirteen "switchbacks." It was a dusty dirt road until the first oil was applied about 1956.

Caretakers. The earliest Balch Park caretaker of definite record was Monroe C. Griggs (1955) who was hired in 1938. He was forced to retire the next year because he was

WISHON POWER PROJECT

seventy-four years old and uninsurable. C. F. Hedrick looked after the park while he had his little sawmill there working up dead and dying trees. This was either just before or just after Griggs. Then Mr. Dalton "Deacon" Clark was caretaker until gasoline rationing during World War II practically eliminated the visiting public. During the war and post-war period the park was maintained by the California Division of Forestry under a cooperative agreement.

Tulare County officials have usually considered Balch Park something of a white elephant. Twice (see Fresno Bee, July 29, 1928 and Porterville Recorder, September 27, 1945), Tulare County came near to donating it to the State, but the Division of Beaches and Parks, after field examinations, turned it down. The restrictive deed from the Balch family was a deterrent.

The park was improved in 1958 and 1959 by construction of fishing ponds and fireplaces. Mr. William Jordan was caretaker in 1959 and 1960. During the summers of 1961 and 1962 Mr. and Mrs. Lloyd Bowdlear took motherly and fatherly care of increasing numbers of campers.

Wishon Power Project. The power development on the Wishon Fork of the Tule deserves more attention. As early as 1905 the San Joaquin Light and Power Company acquired water rights on this stream. Dave L. Wishon, a brother of General Manager A. G. Wishon, was the engineer in charge of surveys and the early construction work. In 1907, a road was built up the Wishon Fork from the Tule Forks where the P. G. & E. power plant now stands. Construction of tunnels to carry the water from a diversion dam near Doyle Springs was begun in 1908.

The project was put into operation January 21, 1914. Water from the natural stream was not considered adequate for dry periods, so it was supplemented by pumping some water into the system from Doyle Springs themselves. After the dry year of 1924 a dam was built across the outlet of Summit Lake at the head of the Wishon Fork. This dam was built by hand tools. About 1930, after a re-survey put the lake inside the Sequoia National Park, the dam was cut through and abandoned.

In 1931 the San Joaquin Light and Power Company sold the controlling interest to the Pacific Power and Light Company. The development since 1936 has been operated under the name of the latter firm. The Wishon Fork water, after passing through the P. G. & E. turbines at the Tule Forks goes into the long flume that carries it down-river to the southern California Edison power plant just above Springville. Then, "after all the electricity has been squeezed out of it," it is piped to Springville for the village water supply.

CHAPTER XVI

THE STATE JOINS IN

> The Mountain Home State Forest shall be developed and maintained pursuant to this article as a multiple-use forest, primarily for public hunting, fishing, and recreation.
>
> Statutes of the State of California,
> Public Resources Code Section 4436.

IN THIS CHAPTER are grouped the happenings that took place on, or with some definite relation to, the privately-owned lands of the North Tule forest area, and therefore have not been included in the two preceding chapters. Just before the final curtain falls on this drama the State makes its appearance as an owner-manager of land.

Among the many pieces of privately-owned land in our four-township area in 1905, three were important enough for special review here. From north to south they are Dillonwood, Camp Lena, and the property first blocked up by the old Tule River Lumber Company and now within the Mountain Home State Forest. A considerable body of information exists concerning the people who owned and used these lands since 1905. That which we deem pertinent to this story and not unwise or unkind to repeat, we shall now recount.

<u>Dillonwood</u>. The 1540 acres we call Dillonwood is probably the largest Sierra redwood property anywhere that is still in private ownership. It consisted in 1905 of two properties. The upper third, mostly virgin forest, was owned by J. M. Canty, the Cantys having purchased several "timber and stone claims" from the original patentees about 1890, in some cases even before the patents were actually signed. It was still a virgin forest of redwood and mixed timber when Canty lost it in 1916 to the state for non-payment of taxes. In 1944 it was purchased by "Doc" Woods, W. J. Sheffler, J. N. Nichols, E. S. Roth, Howard A. Allen, and J. C. Jones, a group of

Southern California business and professional men who wanted a place to hunt and fish. The remaining two-thirds was the N. P. Dillon property which had been acquired by him in various ways (see Chapter IX). After Dillon's death in 1903 it remained undivided in the hands of his many heirs until 1927, their attempts to capitalize on its timber having ended about 1914, as described in Chapter XI. The heirs deeded it to the Bank of Italy (now Bank of America) which in 1944 sold it to "Edgar S. Roth, et al.," probably the same group listed above. The two properties are supposed to have cost the partners about $9,000.

Beginning in 1948 certain parts of the combined property were deeded to the "Dillon Woods Corporation" which was owned by approximately these same people. Jim Lasure and Pat McDonald of Springville report that Ben Steele of Calabasas was the second manager (after the first was caught embezzling the firm's money).

Recent Logging. The Dillon Woods Corporation undertook to make some money from the timber. In 1948 Mal Harris logged some of it to his Springville mill. In 1949 and '50 Pat McDonald (no relation of Jim, Joe, and Earl) was the logger, delivering first to Harris' mill, later to Rauch's. (The Rauch mill, west of Springville, was sold to Harbor Box Company late in 1950.) Pat reports that he cut and hauled twenty million board feet of redwood and five million of pine during those two years, and made more money than in any years before or since. He credits his success to his invention of a tractor-mounted shock-absorbing wedge splitter that he used for splitting the huge redwood logs into sections small enough for hauling on the highway.

Even after forty or fifty years of early-day lumbering on Dillon's property, McDonald found some good timber. He reports cutting a million board feet from sixteen trees standing on a six-acre area. The average of all redwoods cut was 50,000 feet per tree, he says, and the largest 155,000. The present owner estimates that 35% of the redwood was left in the woods because of breakage in falling. Some breakage was pure bad luck. Pat says one tree even broke in the air before hitting the "pillows" prepared for it, leaving only two sixteen-foot logs usable. Pictures of the woods operation appeared in the October 1951 National Geographic Magazine.

Jim Lasure represented the Dillon Woods Corporation when McDonald was logging. In 1951-52 Mr. Carson Maynard logged from Dillonwood to the Harbor Box Company mill, Bud Lyman of Springville representing the land-owners. Prior to 1951 the corporation hired their sawmilling done and sold the lumber, according to Lasure. After that they sold the logs.

In 1957 and '58 logs from the property were sold to Lampe Lumber Company of Tulare, with the woods work contracted to Don Brooks of Porterville, and Al Root of Springville. McDonald estimates the total timber cut from 1948-58 at about 42 million board feet. The highest known stumpage prices for Sierra redwood ($20 to $25 per M.) were obtained by the Dillon Woods Corporation during these years.

Dillonwood Becomes a Tree Farm. Early in 1960, almost a century after the first sawmill went into the "North Tule Pinery," the Dillonwood partners sold their 1540 acres to Forrest Reed of Santa Rosa, an operator of several tree farms in California and Oregon. It thus became the first forest property in this area, and probably in Tulare County, to be owned by a professionally-trained forester. There is reason to hope that this often-destroyed, but so far indestructible, forest will see far better management the second hundred years than the first. The Fresno Bee (February 25 and September 6, 1960) reported the sale consideration to be about $40 per acre, or $62,000, and quoted Reed as estimating that there were a hundred giant redwoods left on the property and four to five million board feet of salvageable waste from the eighty to one hundred million board feet of redwood taken out in earlier logging operations.

Of course the second-growth timber is the property's greatest asset. In a state report (California, 1952) the cutover portion of Dillonwood is described as "one of the finest groves of second-growth Sierra redwood" with trees 100 feet tall and 30 inches in diameter. Reed says in a letter to the writer:

"The redwood will remain the largest single tree in numbers and individual size within the 1540 acres of Dillonwood. It appears very vigorous and shows no sign of diminishing or declining. . . . In my cuttings I am attempting to combine more favorable esthetics, a more vigorous species mixture, and a profitable investment. It may require 20 years to determine whether this can be done but meanwhile I'm doing my best. From the standpoint of history I am sure Dillonwood will survive. In the past it has been influenced; (but) this time I hope improved by its contact with man."

Camp Lena. The eighty acres Jesse Hoskins bought from the government in 1884 to save it from Frasier's sawmill, remained in 1905 almost as God made it. At Hoskins' death in 1908 the property passed to his brothers and sisters. They sold it to Will and Fred Gill in 1913 for $5,000. John Spees of Springville acted as go-between and received his commission "in kind"; namely, one redwood tree. Fred deeded his interest to Will in 1923 (Keagle, 1946). Will Gill is said to have been the first to drive an automobile into the Mountain Home area. The old Frasier grade was so steep that a passenger had to ride on the fender and pour gasoline by hand into the carburetor.

The Gills did not allow any logging until 1956. They did use the dead redwoods as a source of fence posts for their extensive ranching operations. And they permitted Art Griswold to operate a popular camping center and pack station there from 1928 to 1937.

The Pack Station Business. Taking people into the "back country" developed into a sizeable activity by the 1920's. Art Griswold and Frank Negus--who are still ranching in the North Tule area--and the late "Little George" Dillon, all operated packing businesses from their ranches before 1925. Otis Lawson was another local packer. Malvin Duncan of Porterville states that he ran the first pack station at Camp Wishon in 1914. That year he and young Jim McDonald loaded fourteen-foot lumber on miles and packed it up to Jordan Peak for the first lookout house there. George Haigh of Springville says that a Mr. Kirkpatrick operated the Camp Wishon pack station in 1918. Charlie Smith, a son-in-law of old John Nelson, was operating Camp Nelson about that time with Carr Wilson.

Griswold moved to Camp Lena in 1928 as soon as the Balch Park Road was opened. He made packing a full-time job with one to three hired packers. Dillon moved into the old Mountain Home hotel site about two years later. Art has many happy memories of his years at Camp Lena. "Every night there was a big campfire and singing and cutting-up," he says. "We had lots of vacationing school teachers and they would organize the youngsters and give skits and plays and lead real good singing.

THE BIG BLOW

On our pack trips we would take as many as 19 people to a party. Fishing and hunting were the main activities."

The Big Blow. In July, 1933, there was a big windstorm from the northeast. There were thirty-two people camped at Camp Lena that night and thirty-one of them crowded into the soggy 14 x 14-foot room in old Jesse Hoskins' Hercules tree. Dude Sutch says Ed Carter of Porterville, one of Griswold's packers, was the one who insisted on going to bed outside. Trees were crashing down all around. The next morning the road to Shake Camp was blocked by thirty-five windfalls. At Shake Camp a big redwood fell among the campers with no casualties resulting. Griswold was camped that night with a party at Twin Lakes and had "a Hell of a time" getting back through the blowdown. The unofficial fraternity composed of the thirty-one people who spent a long two hours jammed into the Room Tree includes several members of the Morris Gill, Malvin Duncan, and Griswold families. The same storm caused extensive blowdown around Redwood Meadow at Dillonwood.

Another story is told of a redwood that fell on a calm night about 1911 and killed Will Gill's six-horse team stabled in a barn at Camp Lena. Drivers Archie Ainsworth and Skinny Kirk claimed they would have been killed, too, except that they decided that night to move their beds to a new location in the hay loft.

Griswold sold his pack business to R. G. Murdock and Charles Spangler in 1937. Frank Negus and son Roy took over from them in 1941, and established the station at Shake Camp, where they operated it profitably through 1953.

Camp Lena Logged. During the winter of 1955-56 lumberman Mal Harris of Springville looked around as usual for some timber for his next season's operation. He couldn't find anything that he could buy at what he considered a reasonable price until Ralph Gill, who had inherited the management of the Camp Lena "80" after the death of his father, and who is a relative of Harris, agreed to sell him most of the timber on the property. Saved were the historic Room Tree, the Seven Sisters, a few other large redwoods near the buildings, and the few redwoods that were under six feet in diameter.

Harris started falling timber in April 1956. The cutting aroused some objections. A few newspapers and the Sierra

Club (Sierra Club Bulletin, May 1957) deplored the "Havoc in the Big Trees," but no governmental agency was in a position to do anything about purchasing the property on such short notice. The loggers took about six million board feet of timber off the 80 acres that year, most of it redwood. One tree yielded 170,000 board feet of sound wood, enough to build fifty houses. For such a tree Harris' stumpage payment to Gill would have been $1700. Whether Harris made much net profit from the summer's operation is doubtful.

Other Small Ownerships. There were more small owners of timberland in our area in 1905 than now, but no large ranch holdings. Charles Gill did not take over the old Kincaid ranch until about 1915 and had not yet acquired land in the Bear Creek drainage area. The Crook brothers north of Rancheria probably were the largest owners of non-forest land.

There were several small edge-of-the-forest homesteads such as those of Ben Harper (later owned by Hoover, then Charley Elster, now Ed Bace), Clayton Northrop (Ratzlaff's Bear Creek Ranch), John Gaffney at Rancheria, John Lawson, George U. Wray, Jake Abernathy (later owned by Luther Carl), and the Doyle Springs property above Wishon. Several hundred acres of timberland in scattered parcels had been acquired by Rev. F. H. Wales and T. J. Nash of Tulare and the Bank of Tulare. J. J. Doyle also held several properties.

The present Rauch section in the Alder Creek redwood grove was then owned by Samuel L. Davis who had acquired it in 1891 from the original patentee Frank M. Pixley. Back in Peck's Canyon, the Peck's Cabin "160" was in the hands of the original patentee, sheepman Harry Quinn. He sold it to Thomas Pratt about 1907, and it is now owned by Mrs. Frances Pratt of Porterville, his daughter-in-law.

The occupied properties were used for a variety of purposes, mostly coming under the general head of stump-ranching. During the Prohibition years this included a bit of moonshining, but the men of Mammoth Forest who made illegal whiskey shall be as nameless here as the aborigines.

The Hume-Bennett Property. In 1905 William G. Uridge held title to about 2560 acres that he had bought from Greenewald plus several additional properties all of which later was to become part of the Mountain Home State Forest. A small

GOVERNMENT ATTEMPTS PURCHASE

part of it had been logged by Frasier, the Enterprise Lumber Company, and Elster as has been described in Chapter XI. It included Doty's old Mountain Home resort. All of it had been through financial difficulties and more were to come.

The financial failure of the Central California Redwood Company resulted in the sale of this property in 1907 to Ira Bennett, who handled real estate and other business for Thomas R. Hume. Bennett gave Uridge a mortgage for $44,500 and immediately deeded the property to Tom Hume. The Humes were a well-fixed Michigan family whose wealth had been derived from timber and lumber. Bennett originally had a part in the big Hume-Bennett Lumber Company of Fresno County but got out of that sinking ship early, leaving Tom's son George as the managing owner.

In 1920 Tom deeded his Mountain Home land to George. George's financial failure at Hume Lake resulted in his deeding it in 1922 to the Pacific Southwest Trust and Savings Bank (now the Security First National Bank of Los Angeles) which transferred it to the Michigan Trust Company in 1924. This company was usually assumed to be controlled by the Hume family. Tom Hume, before 1920, increased the acreage by purchase of several additional small properties including John J. Doyle's timberland north of Shake Camp, and the Bank of Tulare land in the Rancheria Creek watershed, bringing the acreage to its final total of 4560 acres.

The State was not the Michigan Trust Company's first prospect as a buyer for their forest. It is rumored that they hoped at one time to sell it to Hearst, the newspaper man who had bought an extensive mountain property on the Coast and built a castle on it. But the federal government was the logical purchaser.

<u>The Federal Government Attempts Purchase.</u> The U.S. Forest Service made a strong bid for the Mountain Home Tract. During the depression years of the 1930's the federal government had an active land purchase program. The <u>Fresno Bee</u> in 1936 carried several items about the proposal to acquire the property. For example, on January 7, under the headline, "Hume Prepares To Price Forest For U.S. Purchase," the paper mentioned "George Hume of Detroit representing the Hume interests" conferring with Forest Supervisor Joe Elliott. George Hume was representing the Michigan Trust Company

of Muskegon, Michigan which was the Trustee of the estate of Thomas R. Hume. On January 9 the headline read, "Sequoia Forest Head Advocates Hume Purchase," and Jack Brattin was mentioned along with Hume and Elliott as members of a field inspection party to Mountain Home.

On March 15 the <u>Bee</u> announced: "Elliott Reveals Timber Cruise To Start Soon." Wes Snider says that he and the other Sequoia National Forest rangers cruised the Mountain Home tract in March, working in deep snow from the Mountain Home Guard Station and the Wishon station. But Supervisor Elliott and George Hume were unable to reach an agreement on the price. One reason, according to Norris, was that the government appraisal policies would allow only a very low value for redwood because it was assumed that the government never would sell it for lumber. The redwoods were disregarded in the forest service field cruise, and recreational values were given scant monetary value.

During all the years that Tom and George Hume owned the Mountain Home tract they followed a hands-off policy. People were not prevented from hunting, fishing and camping on it, but no timber was cut. The old lumbering wounds were allowed to heal. The same policy was followed by the Michigan Trust Company until Donald "Dude" Sutch of Springville in 1930 obtained permission to work up dead redwood trees into fence posts. This was the beginning of a long story culminating in acquisition of the Mountain Home tract by the State. But let Mr. Sutch tell it. He was there.

<u>Dude Sutch's Story</u>. "I started in the post-cutting business at Mountain Home in 1929," Mr. Sutch told the writer. "That year I made posts on Gill's Camp Lena property. During the summer, I hiked all over the surrounding Hume-Bennett forest and found lots of redwood windfalls that would eventually rot away if not worked up. Nobody seemed to know who had charge of the property except that Charlie Gill ran cattle on it. Charlie wouldn't give me any information at all, so I went to the assessor to see who paid the taxes. It was Jack Brattin. He lived in Centerville. He said the company had never sold or cut any timber yet, but I got him to come to Mountain Home and meet with me and Wes Snider, the U.S. forest ranger. It was necessary to have Wes in on it because he knew where the

DUDE SUTCH'S STORY

section corners and land lines were and Jack didn't. (Brattin's principal occupation was that of saloon-keeper.)

"I told Jack I wanted all the dead timber or none. Finally he set a price of $2.50 per thousand board feet and said, 'Go ahead until I stop you.' And sixteen years later he still hadn't stopped me.

"At first I worked only on dead trees. And I was the only man allowed to do any cutting at all until 1941. That was the year Claude Rauch came in with his sawmill. (It burned in 1944.) Jack was trying to sell the property. He was asking about $700,000 and the U.S. Forest Service wasn't showing any signs of being willing to meet his price. So Jack figured that some drastic steps were going to be necessary to jar them or the State into action.

"His idea was to mess up the woods by bringing in sawmills. Later he let me and other post-cutters fall green redwoods. The more conspicuous the wreckage was, the better. He figured this would get the public riled up enough to force some public outfit to buy the property. And I guess he figured it about right. The Legislature never would have put up the money of we hadn't made a big noise like we did. You see, the Mountain Home country had been left alone for about thirty-five years and people had gotten to think of it as a virgin forest. Also, the opening of Balch Park had helped to create an interest in keeping the whole area as some sort of a park.

"Before getting sawmills to come in Jack Brattin visited several places where logging was going on. He judged that Rauch's logging methods were ruthless enough to suit his purposes, so he gave him first choice of millsites. I showed them the Hedrick Pond site and the Rauch millsite where the State Forest Headquarters is now. Jack got Mervyn Taylor and Bob Bostrum to put up a little mill at Shake Camp in 1940 or '41. They only had $1200 capital and sawed only a few logs before they went broke. But they ruined Shake Camp. It is a good campground now but nothing to what it was once.

"Another thing that added to the appearance of destruction in the Mountain Home country at this time was the big windstorm that hit the Tule River timber country on January 10, 1941. One of my truck drivers was caught between fallen trees near old Mountain Home and had to walk to Milo. I bought the timber that fell on the road, 150,000 board feet, and had it sawed in Springville. Mal Harris salvaged some pine in 1941 in Sections 34 and 35 and in Section 2 near Crystal Spring.

"Hedrick moved his mill from Balch Park to the Hedrick Pond site about 1939.* And Jack got a man named Brown to cut redwood trees for posts to the south of Balch Park, and Tom Jones to cut trees around Camp Lena and Shake Camp. I worked more to the south. I had as many as 30 men working at times. One year, 1944, we cut and shipped 200,000 fence posts to Valley and Coast points. The highest price was 65¢ for a 7-foot post F.O.B. Springville. During the depression they had gotten as low as 22¢ delivered. In the 1940's Brattin had raised his stumpage price to $4.50 per thousand.

"One tree yielded 12,000 good 4 x 5 posts. The stump stands at the lower end of what you now call Hamer Valley just south of the Gill eighty. It fell straight up the draw. Cost me about $140 for the stumpage.

"I didn't like to see the big trees go down any better than anyone else. But if I hadn't cut them someone else would have. I used dynamite by the ton. If they had left me go another five years there wouldn't have been any good redwood left.

"We fell practically all our trees with dynamite, using two-inch augurs to bore holes for the powder. We would blow a deep undercut a little at a time, then the backcut in one charge. We could fall a tree as accurately that way as with saws and wedges. Only two trees ever fell where we hadn't planned. The first live tree we cut was in 1942 or '43. It was a bad leaner--out over a rock pile. We thought we would put a line on it and pull it over to one side. The problem was to get the line high up in the tree. So we built a cannon. It held half a whiskey glass of black powder and shot a three-and-a-half-pound wooden arrow with 1000 feet of carpenter's cord tied on it. We had a little dog that would wait for each cannon shot and then run and retrieve the arrow. After experimenting and target-shooting with it for a few weeks, we got the cord over the upper part of the tree all right, and then pulled up a light

---

*In 1940 Viro Pickard set up a small mill in Balch Park that he had brought from Oklahoma. His widow Mrs. Dora Pickard of Visalia supplied this information. This mill employed about 8 men and ran 2 years. After the 1941 blowdown Pickard logged some trees on Hume land to the southeast. It is also reported that Buster and Bounce Courtner had a sawmill somewhere on privately owned land in this vicinity during this period.

wire with the cord, then a number nine wire, and finally a five-eighths-inch steel line. We cinched that cable around the top of the tree like a lasso on a steer, using a stump puller to tighten it with. We sawed the tree partly off and pulled for two days with that stump-puller but couldn't budge it. Finally we gave up, blasted it down, and busted it all to pieces on the rocks. So many people were coming up and standing around to watch the job that we were afraid we would break the line and kill somebody.

"After Jack Brattin had his deal with the State almost closed I laid off cutting trees for a month or two. But I couldn't afford that for very long. So I phoned him one morning and said, 'If the State don't sign I start knocking down trees again tomorrow. I've got a ton of powder on hand.' That day they signed. I think it was July 20 or 21, 1945.

"I built the 'House-that-Jack-Built' for Jack Brattin. The lumber came from Rauch's mill. Jack refused to accept the last stumpage payment I owed him because of my help with his sale to the State and my loss in falling thirty-two redwood trees early in 1945 that the State wouldn't let me work up. That last big bang was to force public opinion so the State would sign."

That is the story of the State's purchase of the Mountain Home tract as it looked from the woods. Now let us see what went on in the outside world.

<u>Californians Take Action</u>. If Colonel George W. Stewart may be called the obstetrician responsible for the successful birth of the idea that grew up to be the Sierra and Sequoia national forests; and if John J. Doyle may be called the father of Balch Park; then some such honor is due Arthur H. Drew for his sponsorship of the Mountain Home State Forest. Perhaps "matchmaker" is the term. If it hadn't been for Mr. Drew the eager suitor with the surplus forest might never have won the reluctant lady with the $600,000 dowry.

Mr. Drew was a Fresno attorney and an active member of the Native Sons of the Golden West. It appears likely that his knowledge and interest in the Mountain Home tract came from the Hume interests, as he was a long-time friend of George Hume and acquainted with Strother P. Walton, Hume's Fresno attorney. In any event, Drew came to the conclusion that saving the outstanding redwood groves at Mountain Home at a bargain price to the State would be a worthy project for the Native Sons and Daughters.

The story of the saving of the Mountain Home redwoods might make an exciting movie if it were told in terms of conservationists battling valiantly against "ruthless lumber barons" amidst swinging axes, exploding dynamite and falling trees. The only difficulty is that that wasn't the way it was. The episode might better be described as a campaign to arouse public opinion. Axes, dynamite, and falling redwoods were a part of the strategy, all right, but more sophisticated measures directed through local organizations and the State Legislature brought the final victory.

The published record of the public opinion campaign apparently begins with an item in the <u>Fresno Bee</u> for November 20, 1942, headlined, "Lumber Group Negotiates To Cut Giant Sequoias Surrounding Balch Park." The article stated that Jack Brattin had confirmed that lumber interests, whose names he did not divulge, were negotiating for the timber on the Mountain Home Tract. Whether there was any more to this rumor than a real estate salesman's come-on to quicken the prospective buyer's interest, we have no way of knowing. Goldsmith (1945) says that Walter Johnson, the San Francisco timber and lumber financier, was interested at one time in obtaining the land to trade to the U.S. Forest Service for timber.

This is where Mr. Drew comes in. The <u>Fresno Bee</u> of December 6, 1942 stated, "A move launched by Fresno Parlor No. 25, Native Sons of the Golden West, in an effort to obtain early legislative action to purchase the land. . . . will be presented to the grand parlor this week with a request the purchase plan be brought to the attention of the State Legislature at the 1943 session.

"Arthur H. Drew, Chairman, Herbert McDowell and Lucius Powers, Jr., members of a special committee, informed the lodge the tract may be purchased for $600,000. . . ." The resolution stated that, "During the past season. . . . a small sawmill was erected on the tract and the production of lumber actually commenced. . . . Unless this tract is taken over by the State or some other agency before next season, the logging operations will continue on a larger scale and in a short course of time this magnificent growth of timber will be gone forever."

Two days later the <u>Bee</u> published an appeal by the Fresno Parlor to other organizations to get behind the move. The issue of December 9 reported that Tulare County Supervisors Roy Grogan and R. B. Oliver would back the campaign. The Tulare County Board made no official commitment.

Before December was over the California Federation of Women's Clubs and the Porterville American Legion had gone on record for the proposal. On December 22 Drew, by this time State chairman of the project for the Native Sons, announced in a Bee news item, "We are not asking for another state park, but for a state preserve under the State Division of Forestry, open to vacationists, campers, fishermen, and hunters. Under our plans, state foresters would designate prime lumber which could be cut to keep the forest in its best condition and reduce the fire hazard, as is done in the national forests."

His work soon reached the stage (Fresno Bee, January 23, 1942) that "Senator Hugh M. Burns of Fresno County and Senator Frank W. Mixter of Tulare County late yesterday introduced a bill, SB 339, in the Senate asking for the appropriation of $750,000 for the purchase by the State of the Mountain Home Tract. . . . The tract would become a state forest."

Purchase of the area for a state park instead of a state forest had undoubtedly been the first choice of the sponsors, but the state forest approach was selected as the more likely of the two to gain the support of tax-conscious county supervisors and state legislators. State forests made payments to counties in lieu of taxes, and timber could be sold from them to reimburse the state treasury for the costs of purchase and administration. Besides, the State was concentrating its park purchases at that time in the coastal redwoods country and might have opposed purchase of such a remote and relatively inaccessible area.

The sponsors of SB 339 had no difficulty in getting the endorsement of the Fresno County Chamber of Commerce, but the San Joaquin Valley Natural Resources Committee of the State Chamber failed to agree on it. It appointed a subcommittee to hold a hearing in Visalia on March 25. Drew spoke for the bill, stating that $100,000 to $150,000 could be returned to the state treasury by cutting over-ripe timber, and that, "If we had such a grove in Fresno County, and an effort was made to log it, I would be fighting mad about it." Another hearing was required before the sub-committee voted to urge purchase. It also recommended that the tract be cruised by the State Division of Forestry to determine the fair market price, "taking into consideration the historic and recreational values. . . ."

At Sacramento Drew, Powers, Mrs. C. R. Vanderlip of Coalinga, William Schofield representing the California Forest Protective Association, and Senators Burns and Mixter presented their case before the Senate finance committee early in April, reporting that more than forty organisations and groups had endorsed the purchase plan. The committee listened for an hour, but later turned thumbs down on the bill for reasons that were not made public. This killed it, much to the disappointment of its sponsors.

The next regular Legislature did not convene until two years later, but Drew did not let the matter die. The <u>Fresno Bee</u> of February 6, 1945 stated that Senators Burns and Mixter had introduced Senate Bill 934 to appropriate $600,000 for the purchase of the Mountain Home tract. This was a reduction of $150,000 from the previous bill. Burns was quoted as saying, "More than a hundred huge trees already have been felled." In a later news item about the bill, the <u>Sacramento Bee</u> (June 30, 1945) stated that there were estimated to be only 20,000 giant-sized Sequoia gigantea trees in existence and that 4869 of them were on the Mountain Home tract before felling began in 1941. Goldsmith (1945) later reported that 300 were cut down in the five years prior to purchase by the State.

On May 3, 1945 Drew, as chairman of the action committee, reported victory in sight as a result of a finance committee hearing. The Senate passed the bill on June 8 and the Assembly followed suit on June 16.

California officialdom took little notice of the Mountain Home purchase proposal until SB 934 was passed by the Senate. One exception was a letter written April 30, 1945 from State Forester DeWitt "Swede" Nelson to General Warren T. Hannum, Director of Natural Resources. Mr. Nelson recommended that the tract be acquired for possible trading to the United States Forest Service for land that would consolidate other state forests. Again on May 15 Nelson wrote the Legislative Auditor that the tract "should be in public ownership."

Passage by the State Senate triggered some action by the State Board of Forestry. This board had been officially silent on the proposal previously. It appears that the proponents of the bill, in their enthusiasm, had failed to note that the Board of Forestry was legally responsible for examining and reporting on lands proposed for state forests. The Board had received no request relative to the Mountain Home tract, and had therefore made no determination on it from the standpoint of its suitability for state forest purposes. There were well-founded doubts in the minds of some board members about the legality and workability of SB 934 as written, especially the provision that the tract be preserved "as nearly as possible in virgin state." It is not surprizing that they passed a resolution at their June 15 meeting calling on the Governor to veto the bill.

Governor Earl Warren discussed the bill on July 9 with Director Hannum and State Forester Nelson. He requested that Mr. Nelson examine the tract and report back to him. Nelson immediately went to Tulare County and, in company with Deputy State Forester Cecil E. Metcalf, looked over the tract. In his report to the Governor dated July 13, Nelson expressed himself as "greatly disturbed" to find fence-post cutting still in progress in at least six locations on the tract; and "shocked to find logging operations underway" at Hedrick's mill and on the Rauch cutting area. He was "impressed by what appeared to be forcing of public sentiment for public acquisition. . . . by the cutting operations. . . ." He also stated that when he discussed the situation on the previous day with "General Ray Hayes, former Adjutant General, who is one of the proponents of the bill. . . . General Hayes was dumbfounded to learn that operations were still underway, because he had been told that no cutting would be carried on pending action by the State."

The State Forester, in making his report to the Governor, found himself in the difficult position of being forced to choose between his obligations to the Board of Forestry and his convictions as a result of seeing the possibilities in the Mountain Home tract and the mis-directed operations taking place there. In the interests of the people of the State, he and Mr. Metcalf could see no other way than to do what they could to preserve the tract and its giant trees for the public. Mr. Nelson, therefore, recommended that the Governor sign the bill, and then wrote a letter to the Board of Forestry explaining his reasons. After the bill became law the Board visited the area many times and became extremely interested in its possibilities as a state forest for demonstrating the compatibility of recreation uses with timber growing and harvesting.

On July 19, 1945, Governor Warren signed the bill. Two days later the Michigan Trust Company gave the State a 90-day option to buy, and sent Rauch and Hedrick notices to terminate their logging operations within 30 days. Sutch was notified not to "open up" any more of his felled redwood trees.

There still remained the matter of settling on a price for the tract. The State hired C. B. Goldsmith, a retired U.S. Forest Service forester, to appraise it. "Goldy" spent two months cruising and mapping the tract and checking previous cruise figures furnished by the owners. He, and on occasion State Forester Nelson and Deputy Metcalf, boarded with Dude Sutch and wife at the old Mountain Home cookhouse. Cordial relations prevailed now that the smoke and din of battle were past. Goldsmith (1945) reported that the company's claim of 244,365,000 board feet—over half of it redwood—was an acceptable basic figure, but that 38 million had been logged or made into posts in recent years and another 34 million was inaccessible for economic timber management. This left 172 million as the merchantable accessible stand, of which 79 million was redwood. A price of $550,000 was finally agreed upon and the deed was signed January 6, 1946.

During 1946 the State Division of Forestry set about the job of putting the new forest under management. In doing so it came up against the provision in the law to which the Board of Forestry had objected, the "as-nearly-as-possible-in-virgin-state" clause, which made it doubtful whether live timber could legally be sold, even for clean-up purposes. To clarify this point, Senator J. Howard Williams of Porterville in March 1957 introduced Senate Bill 605. Williams, always a staunch friend of the Mountain Home State Forest, also introduced a bill to permit land exchanges to block up the Forest. Both provisions were enacted into law as Section 4436 of the Public Resources Code.

Cecil Metcalf lost no time in attacking the herculean clean-up job bequeathed him by Mr. Brattin. The <u>Fresno Bee</u> of March 31, 1946 announced that the Division of Forestry was calling for bids for the sale of the usable down timber. Metcalf had already put all of his available men and bulldozers to cleaning up the worst of the wreckage, and had taken steps toward setting up a long-range forest management program. But now we are getting beyond the scope of this story.

So it was that Jack Brattin's thunderous assault in the woods, plus the constructive work of Arthur H. Drew and many others, finally resulted in a state forest with special recreational features and legal authorities.

CHAPTER XVII

TO SIT . . . . TO MUSE . . . . TO SLOWLY TRACE

> To sit on rocks, to muse o'er flood and fell,
> To slowly trace the forest's shady scene,
> Where things that own not man's dominion dwell,
> And mortal foot hath ne'er or rarely been;
> To climb the trackless mountain all unseen,
> With the wild flock that never needs a fold;
> Alone o'er steeps and foamy falls to lean;
> This is not solitude; 'tis but to hold
> Converse with Nature's charms, and view her stores
>   unrolled.
> <div align="right">Byron, <i>Childe Harold's Pilgrimage.</i></div>

THE TRAVELER who must find his own way across an unfamiliar land pauses occasionally to study the terrain. The reader who has plodded up through the preceding chapters to the vantage point of the present, has earned the right to sit a while and rest. But like a seasoned and prudent explorer, he naturally wants, "while he is resting," to review the back trail and study the country ahead. Therefore, let us "slowly trace the forest's shady scene" through the past century; and then peer briefly ahead before we "climb the trackless mountain" of the future.

If the lessons of history teach us anything at all they show that PEOPLE put PRESSURE on RESOURCES. Fortunately for mankind over the long pull, forests are renewable resources that can push back--or even spring forth as strong as ever if people turn their backs a while.

A hundred years ago the balance of power between man and the forest was in favor of the forest. No matter how hard the early white settler might fight it, the forest's renewing power protected it from destruction. But gradually man's power has increased until now he holds forces in leash that can destroy not only the forest, but the world in which it is rooted. The wonder is that after a century of sheep and cattle, logging and post-cutting, sawmilling and road-building, anything has survived. How has it happened that much of the Mammoth Forest is still the same "wooden wilderness" that it was when the first white man—even the first Indian—saw it? How is it that most of the acreage remains untouched by ax or bulldozer? Why do thousands of mammoth Sierra redwoods still stand in confident security from all save winter snows and summer lightning? And why is it that where the old forests were cut down, thrifty new ones have been allowed to grow unmolested so that the virgin forest and the re-growth are sometimes hard to tell apart?

Let's look at the record—broadly. First the Indian. Then the Indians plus a few unknown trappers and prospectors; perhaps a roving sheepman or two. These people lived with the forest, taking little that the land could not readily replace.

Soon the white invaders eliminated the Indian, who had harvested acorns and hundreds of other kinds of wild seeds; and replaced his foraging with that of hogs and sheep. Sheep might still be harvesting some of the forest's plenty if men had not gone greedy and permitted a cloven-hoofed footrace every spring for the first spears of grass on the south slopes. The soil erosion that resulted from this feverish rush for free forage over a period of more than forty years was, after the liquidation of the Indian, the first great tragedy in the mountains. No doubt, fishermen still suffer the results of it in the ripped-out stream channels that follow each unusual rainstorm.

In contrast to the sheepmen, the settlers of this period with their hogs and cattle, even though there were more of them than today, made no great impact on the forest. Neither did the many hunters and other recreationists, nor the prospectors. What the sheep did to the mountains was what awakened the public. Public opinion has been the force that has stood between the forest—and its soil and water—and the pressure of ever more powerful, but scarcely more altruistic, mankind.

Close on the heels of the first domestic livestock came the early manufacturers of lumber and fence material: Hubbs, Wilson, Jim Kincaid, McKiearnan. Their thirty-five years of labor made hardly a dent in the "inexhaustible" timber resource. Geography and economics were on the side of the forest. But with the passage of the Timber and Stone Act in 1878, anyone with the gambling urge and $400 could set out to seek his fortune in the western woods.

TO SIT

"Small business men" came with big ideas. Nate Dillon, the Elsters, Haughton, Avon Coburn, Frasier, the Dotys, the Youngs, Hiram Brey, and John J. Doyle planned, labored, and hung on; but eventually failed to make it pay. They retired or went elsewhere to try their luck. Dillon, the pants-pocket banker and merchant of standing timber, was somewhat exceptional. So were Coburn, the lumberman, and Doyle, the "public relations man," because they carried on successfully from the beginning until at least the latter years of the period.

When the bigger boys came along fortified by corporation papers, they just went broke in a bigger way. Frank Boole, who in the Converse Basin logged every standing stick except his namesake tree, couldn't--or at least didn't--lick the North Fork of the Tule. Comstock, Uridge, the Nofzigers, and even Tom Hume tried their hands, but high mountains and low prices protected the forest yet a while. None of these men were much different than ourselves. If they had had more capital, or better tools or higher prices they might have succeeded in cutting ten times as many trees. Public opinion, then as now, had little effect on what a man did on his own land.

Recreation seekers quickly followed in the footsteps of the first lumberman-road builders. By 1905 thousands of people had worked, roamed, and lived in the big-tree woods, had learned dozens of redwood trees by their given names, and had danced the Virginia reel at Doty's for Saturday nights uncounted. Then came a quieter period; but recreation has again become increasingly important. The trend has been away from long camping trips, often by packstring, toward shorter visits by automobile or two-wheeled vehicles.

It probably was the physiography of the North Tule forest country, as well as the economic conditions of the early days, that protected the redwoods from destructive logging; but chance had a lot to do with it. This area was one of the first to be exploited. Had circumstances held off the attack until a little later, many more of the red giants might have felt the axe.

The failure of the mountains to yield consistent profits, plus the growing public dissatisfaction with their unregulated use and abuse, brought a trend back to public ownership. By 1905 all of the lumbermen were gone from the Mountain Home area, and only a few years remained at Dillonwood. The Sequoia National Forest set out to control grazing and fires, and to manage its immense timber and recreation resources. Balch Park and Mountain Home State Forest eventually were established on land formerly privately-owned.

By 1940 improved economic conditions, better roads, the chain saw, the bulldozer, and the modern logging truck brought logging and fence-post cutting back again. On private lands these operations were no improvement over those of fifty years earlier until the recent change in the ownership of Dillonwood. But on federal and state lands logging practices were greatly improved and a ban put on the cutting of the large Sierra redwoods.

And now, what does the jumbled topography of the future hold in store for us? It is good to look back once in a while, if only because a review of difficulties overcome gives us courage for another climb. But the climb is the thing. To press forward is our heritage and our principal business.

It was a calculated risk for this writer to try to search out and interpret the past; but it is downright foolhardy to predict the future. Nevertheless, the trail leads that way and we follow it. One conclusion seems safe. We have not reached any summit in the management of mountain lands such as those of the North Tule. We may have gotten safely past the sheep problem and the crisis of unregulated and unlimited tree-cutting. But there are and will be other dangers and problems.

The dangers of the future, we venture to predict, will be three: the danger of destruction by fire; the danger of rushing headlong into the conversion of "wild" old-growth forests into managed forests without enough careful planning and enough tested knowledge and tools to do it properly; and the danger generated by public pressure for recreation areas.

How to minimize the fire danger is engaging the attention of many able men both in and outside of public employ. We leave the problem with them, and God speed its solution!

How to harvest the trees that should be harvested—the pine, fir, cedar, and even the second-growth redwood—by methods that will preserve what should be preserved, is a problem on the minds of comparatively few. Taking land out of private ownership and making it public property does not change human nature. The never-ending economic urge to convert standing timber into fluid dollars is still there. Of course, good forestry requires that trees be cut when economically or biologically mature. If man does not harvest the trees, insects and diseases will; and this harvesting method compounds the fire hazard. But there are places where all of the money value of the timber harvest might well be "plowed back" to protect the soil and recreational values, and to insure the next tree crop.

The problems are many. For only one example, should the fir and pine around the feet of the giant redwoods be harvested and thinned for useful timber products and for open vistas for the sightseer; or should they be left to grow so thick and tall that, to transpose, an old saying, "you can't see the trees for the forest"? How to quickly reproduce stands of timber after cutting is another problem that has not been solved in the southern Sierra. The roots of forestry—its scientific foundation—are long. They reach deep in time and draw heavily on the basic sciences. But its visible stem and crown—the art and practice of forestry—have much growing yet to do, and in this new country, showy flowers are few and far between. To bring forestry to its proper stature and fruitfulness will require firm and skillful hands on the pruning knives.

But perhaps the greatest present danger is from recreation pressure. Is it progress to have preserved the forest from the dusty bands of sheep and the slashings of itinerant choppers and sawyers, only to find the same noises, smells, and beer cans as along Highway 99? For what shall it profit a forest, if it shall gain the whole human world, and lose its own soul? It was because of public opinion and pressure that most of the Mammoth Forest country is now in public ownership. To complete the job they have started, the public must see to it that recreation is soundly planned and looked after; so that the use of the mountains will be kept in balance with the resources and the facilities currently available. It may even be necessary to ration mountain recreation to save any of it for our children, as water is rationed among wayfarers crossing a thirsty desert.

And now, friends, it has been a long trail; but we have traveled with interesting people. Meeting them where they worked and played has been a rare pleasure for the writer. He hopes that you, too, have enjoyed passing the time of day with the men of the Mammoth Forest.

The End

REFERENCES

ANDREWS, Ralph W. 1954. This Was Logging. 157 p. illus. Superior Publ. Co., Seattle, Wash.

BANCROFT, H. H. 1884-90. History of California. 7 vol. The History Co., San Francisco.

BARKER, John. 1955. San Joaquin Vignettes. Edited by W. H. Boyd and G. J. Rodgers. 111 p. illus. Kern County Hist. Soc., Bakersfield, Calif.

BARTON, H. 1907. "Lumbering in Tulare Co." typewritten. 21 p. Calif. Div. of Forestry files, Fresno, Calif.

BARTON, Stephen. 1874. "Early History of Tulare County." Visalia Weekly Delta, July 2 - Nov. 26, 1874.

BREWER, W. H. 1930. Up and down California 1860-64. Yale Univ. Press, New Haven, Conn.

BROWN, Mrs. Jay. 1923. "The Centennial Stump." Daily Recorder (Porterville, Calif.), May 8, 1923.

BURCHAM, L. T. 1959. "Planned Burning as a Management Practice for California." mimeographed. 17 p. Calif. Div. of Forestry, Sacramento.

CALIFORNIA, STATE OF. 1952. "The Status of Sequoia gigantea in the Sierra Nevada." mimeographed. 75 p. illus. maps. Report to the California Legislature.

CAUGHEY, John Walton. 1940. California. 680 p. Prentice-Hall, New York.

CHALFANT, W. A. 1922. The Story of Inyo. 358 p. maps. Hammond Press, Chicago. (Revised 1933).

REFERENCES

CLAR, C. Raymond. 1959. California Government and Forestry. 623 p. State of Calif., Sacramento.

CLELAND, Robert Glass. 1922. A History of California: the American Period. 512 p. The Macmillan Co., New York.

_____ 1951. The Cattle on a Thousand Hills, 1850-1880. Second edition. 365 p. The Huntington Library, San Marino, Calif.

COOK, S. F. 1955. Aboriginal Population of the San Joaquin Valley. 5 maps. Univ. of Calif. Press, Berkeley.

CREEL, George. 1926. "Incredible Kit Carson." Collier's: The National Weekly, Aug. 21, 1926.

CUTTER, Donald C. 1950. "Spanish Exploration of California's Central Valley." typewritten. 275 p. map. Ph.D. dissertation. Univ. of Calif. Library, Berkeley.

DANA, Samuel T. and Myron KRUEGER. 1958. California Lands. 308 p. American Forestry Ass'n., Washington, D. C.

DASMANN, William P. 1958. Big Game of California. 56 p. Calif. State Dept. of Fish and Game, Sacramento.

DERBY, George H. 1852. Report of the Secretary of War Communicating . . . . a Report of the Tulare Valley Made by Lt. Derby. 16 p. map. Washington, Govt. Printing Office.

DOCTOR, Joseph E. 1959. Historical articles in Centennial Edition of Visalia Times-Delta, June 29, 1959. 9 sections, 144 p. Visalia, Calif.

DOTY, Charles. 1947. Letter to H. G. Schutt. 9 p. Tulare Co. Historical Society files, Visalia, Calif.

DOUGLASS, A. E. 1945-6. "Survey of Sequoia Studies." Tree Ring Bulletin - 11 (4): 26-32, 12 (2): 10-16, and 13 (1): 2-8, 5-6. Laboratory of Tree-Ring Research. Univ. of Arizona, Tucson.

DUDLEY, Wm. R. 1896. "Forest Reservations: with a Report on the Sierra Reservation, California." Sierra Club Bulletin 1 (7): 254-267. San Francisco, Calif.

_____ 1899. "Forestry Notes." Sierra Club Bulletin 2 (5): 292. San Francisco, Calif.

_____ 1900. "Forestry Notes." Sierra Club Bulletin 3 (1): 118, 186. San Francisco, Calif.

DYER, Hubert. 1898. "The Mt. Whitney Trail." Sierra Club Bulletin 1 (1): 1-8. San Francisco, Calif.

ELLIOTT, Wallace W. & Co. 1883. History of Tulare County, California. . . . Publ. by Wallace W. Elliott & Co., San Francisco, Calif.

ELSASSER, A. B. 1962. Indians of Sequoia and Kings Canyon National Parks. 59 p. Sequoia Natural History Association, Three Rivers, Calif.

FRY, Walter. 1930. Big Trees. 114 p. Stanford Univ. Press, Stanford, Calif.

GOLDSMITH, B. C. 1945. Two memos, 10 p., Oct. 1, and one report, 17 p., Oct. 2, 1945 to State Forester. typewritten. Mountain Home State Forest files, Springville, Calif.

GREELEY, Wm. B. 1951. Forests and Men. 255 p. Doubleday & Co., Garden City, N. Y.

GRIGGS, Monroe C. 1955. Wheelers, Pointers, and Leaders. 60 p. Tulare County Hist. Soc., Visalia, Calif.

GUDDE, E. G. 1960. California Place Names. 383 p. Univ. of Calif. Press, Berkeley.

GUINN, J. M. 1905. History of the State of California and Biographical Record of the San Joaquin Valley. 1643 p. Chapman Publ. Co., Chicago.

REFERENCES

HARTESVELDT, R. J. 1962. "The Effects of Human Impact upon Sequoia gigantea and its Environment in the Mariposa Grove, Yosemite National Park, California," Ph.D. dissertation. mimeo. Univ. of Michigan, Ann Arbor.

HOLBROOK, Stewart H. 1938. Holy Old Mackinaw. 278 p. The Macmillan Co., New York.

HURT, Bert. 1941. "A Sawmill History of the Sierra National Forest, 1852-1940." 50 p. mimeographed. illus. U.S. Forest Service, Fresno, Calif.

KEAGLE, Cora L. 1946. "Early Day Lindsay Man Made His Home in Giant Sequoia." Fresno Bee, Dec. 16, 1946. Fresno, Calif.

KING, Clarence. 1902. Mountaineering in the Sierra Nevada. 378 p. Scribner's and Sons, New York.

KROEBER, A. L. 1953. Handbook of the Indians of California. 995 p. Calif. Book Co., Berkeley.

LATTA, F. F. 1934. "Little Journeys in the San Joaquin." illus. Clippings from newspapers. Tulare County Library, Visalia, Calif.

_____ 1936. El Camino Viejo a Los Angeles. Kern County Hist. Soc., Bakersfield, Calif.

_____ 1949. Handbook of Yokuts Indians. 287 p. illus. Bear State Books, Oildale, Calif.

LEADER, H. A. 1928. "The Hudson's Bay Company in California." Unpublished thesis. Univ. of Calif., Berkeley.

LEWIS PUBLISHING CO. 1892. Pen Pictures from the Garden of the World, Memorial and Biographical History of Central California, Counties of Fresno, Tulare and Kern.

LO JACONO, S. 1959. Establishment and Modification of National Forest Boundaries, a Chronological Record,

1891 - 1959. U.S.D.A. Forest Service, Division of Engineering, Washington, D. C.

LONGHURST, W. A., Aldo S. LEOPOLD, and R. F. DASMANN. 1952. A Survey of California Deer Herds. Game Bulletin No. 6. 144 p. Univ. of Calif., Berkeley.

MATTHES, Francois E. 1930. Geologic History of the Yosemite Valley. Professional Paper 160. U.S. Geological Survey.

MAYFIELD, Thomas Jefferson. 1929. Uncle Jeff's Story. Arr. by F. F. Latta. 88 p. Tulare Times, Tulare, Calif.

McCULLOCH, Walter F. 1958. Woods Words. 219 p. Oregon Hist. Soc. and the Champoeg Press, Portland, Oreg.

MCGEE, Lizzie. 1952. "Mills of the Sequoias." mimeographed. 27 p. Tulare County Hist. Soc., Visalia, Calif.

MEEUWIG, R. O. 1960. "Watersheds A and B - A study of Surface Run-off and Erosion in the Subalpine Zone of Central Utah." Journal of Forestry 58 (7): 556-60.

MENEFEE, E. L., and F. A. DODGE. 1913. History of Tulare and Kings Counties, with Biographical Sketches. 890 p. illus. Historical Record Co., Los Angeles, Calif.

MITCHELL, Annie R. 1962. Letter based on unpublished material in Tulare County Historical Society files. Visalia, Calif.

MORGAN, George T., Jr. 1961. William B. Greeley, a Practical Forester. 82 p. illus. Forest History Society, St. Paul, Minn.

MUIR, John. 1876. "God's First Temples; How Shall We Preserve Our Forests?" Sacramento Record-Union. February 5, 1876. Sacramento, Calif.

_____ 1878. "The New Sequoia Forests of California." illus. Harpers Monthly. 57 (Nov.): 813-827.

REFERENCES 151

_____ 1894. Mountains of California. 381 p. The Century Co., New York.

_____ 1901. "Hunting big redwoods." Atlantic Monthly. 88 (Sept.): 304-320.

_____ 1938. John of the Mountains. 459 p. Houghton Mifflin Co., Boston.

MUNOZ, Fray Pedro. 1806. "The Gabriel Moraga Expedition of 1806: The Diary of Fray Pedro Munoz." The Huntington Library Quarterly, IX: 223-28, 1945-6.

NATIVE DAUGHTERS OF THE GOLDEN WEST. 1954. Old Cemeteries of Southeastern Tulare County. 18 p. illus. The Farm Tribune, Porterville, Calif.

NORBOE, P. M. 1903. "Trails into the Mt. Whitney and Kern River regions." Mt. Whitney Club Journal, I (2): 60-71.

ROBINSON, W. W. 1952. The Story of Tulare County and Visalia. 36 p. Title Insurance and Trust Co., Los Angeles, Calif.

RENSCH, H. E. and E. G., and M. B. HOOVER. 1933. Historic Spots in California. 597 p. Stanford Univ. Press, Stanford, Calif.

SCHULMAN, Edward. 1956. Dendroclimatic Changes in Semiarid America. 142 p. Univ. of Arizona Press, Tucson.

SCHUTT, Harold G. 1962. "Prehistoric Rock Basins." Los Tulares No. 54. Tulare County Historical Society, Visalia, Calif.

SMALL, Kathleen Edwards. 1926. History of Tulare County, California. 2 v. 504 p. S. J. Clarke Publ. Co., Chicago.

SMITH, Wallace. 1939. Garden of the Sun. 558 p. Lymanhouse, Los Angeles.

STEWART, George W. 1929. "Prehistoric Rock Basins." American Anthropologist, July - Sept., 1929.

_____ 1933. "Tulare County History Scrapbook." Clippings dated between 1900 and 1928. 166 p. Donated to Tulare County Library, 1933. Visalia, Calif.

STINER, Ina H. et al., 1934. "History of Porterville," typewritten. 625 p. Porterville Public Library, Porterville, California.

_____ 1956. "An Appreciation of the Past." typewritten. 2 vols. 482 p. illus. Porterville City Library, Porterville, Calif.

TAYLOR, Ron. 1960. "Rain Caused Sierra Avalanche of 1867." Fresno Bee Jan. 17, 1960. Fresno, Calif.

THOMPSON, Fred. 1892. Thompson's Atlas of Tulare County.

TOWNE, Charles W. and Edward W. WENTWORTH. 1945. Shepherd's Empire. Univ. of Oklahoma Press, Norman, Okla.

TREADWELL, Edward F. 1931. The Cattle King. 307 p. The Macmillan Co., New York.

TULARE COUNTY CHAMBER OF COMMERCE. 1959. A Few Facts About Tulare County.

TULARE COUNTY HISTORICAL SOCIETY. 1949. Historical Bulletin No. 4. 8 p. Tul. Co. Hist. Soc., Visalia, Calif.

_____ 1950. Historical Bulletin No. 6. 8 p. Tul. Co. Hist. Soc., Visalia, Calif.

_____ 1955. Los Tulares No. 24. 4 p. Tul. Co. Hist. Soc., Visalia, Calif.

_____ 1956. Los Tulares No. 26. 4 p. Tul. Co. Hist. Soc., Visalia, Calif.

REFERENCES

_____ 1957. Los Tulares No. 32. 4 p. Tul. Co. Hist. Soc., Visalia, Calif.

_____ 1958. Los Tulares No. 38. 4 p. Tul. Co. Hist. Soc., Visalia, Calif.

WALKER, Frank J. 1890. "The Sequoia Forests of the Sierra Nevada - Their Location and Area." Zoe, A Biological Journal 1 (1). Sept., 1890. San Francisco.

WENTWORTH, Edward N. 1948. America's Sheep Trails. 667 p. illus. maps. Iowa State College Press, Ames.

WILKINS, Thurman. 1958. Clarence King, a Biography. 441 p. illus. The Macmillan Co., New York.

Appendix A

ORIGIN OF PLACE NAMES

To paraphrase Abraham Lincoln's statement about the common people, God must love the little out-of-the-way places because he made so many of them. And to someone, each little brook, hill, and turn of the road has its name.

The following names are found on maps or are in general use locally.

Arrastre Canyon: For the remains of a Spanish "arrastre," a mule-power ore-grinding mill, found in that area. (It drains into the place called "Packsaddle" north of Long Meadow.)

Backbone Creek: The Dillon Mill Road had, and still has, a half-mile stretch that followed the crest of a very narrow, sharp, low ridge between the North Fork of the Tule and a small creek that paralleled it. This piece of road was called "The Devil's Backbone." The adjacent creek was named from it.

Bear Creek: Probably the oldest named geographic feature in the Tule River drainage-area, excepting the Tule River itself. Was the only creek shown by name on government survey plats of the four townships in the North Tule forest area, excepting "Mill Creek" (now called Rancheria Creek). Possibly Bear Creek, like Deer Creek further south, kept its Indian name, merely being translated into English. If so, its Indian name, according to Don Witt, would have been N'HAWN TUL-TUH (Bear Canyon).

Brownie Meadow: There was a shake-maker known only as "Brownie" who worked near this meadow in the 1890's according to I. Elster and Mrs. Ola Hubbs. Whether he or the meadow was named first, we cannot be sure. It is also said that the

APPENDIX A   155

meadow was named for Clinton Brown, a pioneer sheepman of the area.

Camp Lena: A. B. Tienken of Lindsay says, "Camp Lena was named after Mrs. E. Vande Bogart--maiden name Lena Millinghausen--(daughter of the late August Millinghausen of Lindsay and Springville). A group including the Millinghausens often spent their summers at the Mountain Home and Summer Home area and Lena was always the life of the camp, leading the evening entertainments." One night a group at Hoskins' place "christened Camp Lena after Lena Millinghausen."

Cramer Creek or Kramer Creek: Maps show this creek that drains off Blue Ridge into the North Fork, spelled with a "K." However, it was named for Jacob Cramer, a first settler in the area. He had a daughter, Belle, who married John Hossack, a Porterville sheepman and barkeeper (for whom Hossack Meadow and Hossack Creek are named). Mrs. Harriet Maxey of Redding, one of Belle Hossack's granddaughters writes, "When the CCC camp went in above Springville (about 1934), they made the horrible mistake of spelling it Camp Kramer. I can still see that poor colonel trying to extend the government's apology to my grandmother. Her rocking chair was clocking miles a minute."

Dennison Peak: Don Witt says that Dennison was a bear hunter who rigged up a "set-gun" for a bear near the old water-power sawmill above Jack Flats. Then he was careless enough to walk into his own trap, and trip the wire that discharged the gun. When his body was found it was buried where he died; and the Dennison Trail, and Dennison Peak, Mountain, Ridge, Ditch, and School District inherited his name. This must have happened very early, probably in the 1850's or '60's. If he had a given name or initials they have been forgotten.

Frasier Mill: For L. B. Frasier, who built a road to Mountain Home in 1885 and built a sawmill at this site the same year. (Frazier Valley is spelled differently because it was named for another man.)

Galena Creek and Silver Creek: Old timers say that originally the southern of these two streams was called Galena

Creek and the northern one Silver Creek, the former named for the lead (galena) ore found in its drainage-area, and the latter for similar reasons. All modern maps, however, show Galena Creek north and Silver Creek south, and it seems better to accept what appears to be a map-maker's error rather than try to correct it.

Hedrick Pond: For C. F. Hedrick, sawmill man of Lindsay, who originally built it as a mill pond about 1940.

Jack Flats: Shown as Jack Ass Flats on Dillon Road survey notes of 1877. Undoubtedly this and nearby Jenny Creek were named for Nate Dillon's miles that pulled the empty lumber cars from the lumber dump back up his wooden railroad to the mill.

Jordan Peak: Named for John J. Jordan, pioneer trail builder, whose trail went over or just south of this peak and on to the Owens Lake country.

Lumreau Mountain: For Charles M. Lumereau, whose pioneer ranch was at the western foot of this high brushy hill. He spelled his name Lumereau, not Lumereaux, Lumreau, Lumro, or Lumbro; but the mountain is Lumreau on all modern maps.

Maggie, Mount and Mount Moses: There is usually more than one story about how places got their names. Although the books on California names have their version, a composite of somewhat varying stories current among members of the Fred Wells and Kincaid families is that one time when the Kincaids were on a hunting trip and camped near Mount Maggie, a government surveyor, more or less lost, dropped into camp and stayed with them. His name was Moses Peabody. He praised Maggie Kincaid's biscuits to the skies and asked how she would like to have a mountain named for her. "Well," she said, "today I sat right on top of this mountain and it doesn't have a name." One thing led to another until one of the peaks of the highest group in the Tule watershed became Mount Maggie and its craggy companion across the canyon to the west took the given name of the surveyor. (Mount Maggie is not the highest point on Maggie Ridge. The highest points are two nicely

APPENDIX A

rounded twin peaks locally known by a descriptive name that map-makers do not put on maps except in French or Indian).

McDonald Hill: For James McDonald, early-day owner of the homestead just below the ranch that has recently been known as the Cypert Turkey Ranch and the Greer place. The hill is southwest of the Rancheria Ranch and northwest of the "Scicon" Camp.

Milo: Don Witt of Porterville says that his Aunt Agatha Richardson was delegated to make up a list of names for the new post-office for the North Tule community. All were invited to suggest names including her children. The children's choice was Milo, which was what they called their dog. Aunt Agatha, to please them, added Milo at the bottom of her list and the Post Office Department chose it in preference to all the carefully considered proposals of the elder citizens.

Shake Camp: Harry Wells of Bakersfield says his father, Fred Wells, was the first man to split "shakes" at this place. It was in 1891. They were split with a "froe" from sugar pine trees for roofing.

Sycamore Creek: This, the first tributary of the North Fork above Bear Creek, was probably named for the "Sycamore Camp or Spring" mentioned in an 1870 petition for a toll road. This old camp was apparently a principal stopping place on the old Dennison Trail between the Lindsay vicinity and Mountain Home, and was located between the North Fork and Greer's Sycamore Creek Ranch. It may be one of oldest surviving place names in our area.

Tule: "The word is derived from tullin or tollin. . . . it designates the cattail or similar plants . . . . Tule River . . . . appears as 'Tule River or Rio San Pedro' on Derby's map of 1850." (Gudde, 1949).

Watermill Ridge: (South of Dillonwood tract and east of Jack Flats). From the old waterpower sawmill in Section 9, thought to have been built by J. R. Hubbs.

Wishon Fork and Camp Wishon: Named for the Wishon family who were connected with the San Joaquin Power and Light Company. A. G. Wishon was General Manager and his brother Dave L. had charge of the water-development surveys and early construction. Dave's ashes are at the cabin site at the Tule River headworks.

Appendix B

MEMORABLE DATES

Following is a summary "for the record" of notable natural events and "Acts of God" that are mentioned in livestock histories, weather reports, river run-off records, and other sources:

1828-30 Twenty-two-month drouth in California, 40,000 cattle and horses died (Wentworth, 1948).
1840-41 Another drouth.
1849-50 Excessive rains (Elliott, 1883).
1850-51 Dry period (Schulman, 1956).
1852-53 Flood year (Elliott, 1883). Tulare Lake as high as in 1862 (Thompson, 1892).
1855 One of Visalia's memorable floods (Tul. Co. C. of C., 1959), and a severe earthquake (Doctor, 1959).
1855-56 Extraordinary drouth (Elliott, 1883), 100,000 cattle died in Southern California.
1856 Coast Range and San Joaquin earthquake (Barton, 1874).
1860-61 Dry.
1861-62 The all-time "high-water year" in the San Joaquin Valley (Brewer, 1930). Precipitation 215 per cent of normal in Central Sierras (Schulman, 1956). Tulare Lake covered 760 square miles (Thompson, 1892).
1862-64 "The Great Drouth." A million cattle died in California (Treadwell, 1931). "Perfect devastation" (Brewer, 1930). End of cattle boom (Cleland, 1951).
1867-68 The great landslides and landslide floods. Precipitation twice normal in Central Sierras.
1870-71 Drouth.
1872 (March 26). Lone Pine earthquake, "perhaps the greatest California earthquake on record" (Doctor, 1959), felt throughout the southern Sierras.

1872-73 Most severe winter ever experienced by sheepmen (Menefee and Dodge, 1913).

1873 Tulare Lake within 3-1/2 feet of its 1862 peak (Thompson, 1892).

1876-77 Very dry years (Montpelier, 1884), marking end of sheep boom (Cleland, 1951).

1877 Severe drouth. "Sheep and cattle driven into mountains starved to death" (Dasmann, 1958).

1878-79 Dry (Cleland, 1951).

1879-80 A very hard winter, extremely deep snow (Longhurst, et al., 1952).

1881-83 Drouth years in the San Joaquin.

1887 "Terrible rain storm" in August in Mountain Home.

1889-90 Floods. Precipitation 186 per cent of normal in central Sierras (Schulman, 1956). Most severe winter of record, especially hard on the deer (Longhurst et al., 1952), probably an all-time low for deer population.

1892-93 Severe winter (Dasmann, 1958). Hard on deer.

1894-95 Severe winter (Dasmann, 1958). Hard on deer.

1897-99 Drouth years. In July, 1898, Tulare Lake went dry for the first time (Latta, 1934).

1899 A bad forest fire year. Dudley (1900) reported that 850,000 acres burned north of the Tehachapis.

1901 A high water year at Tulare Lake (Latta, 1934).

1905-06 Flood year on Kern River and in Visalia (Tul. Co. Hist. Soc., 1956). At Giant Forest the snow was 29 feet deep, with 12 feet still left on July 4 (Fry, 1930).

1906-07 Severe winter. Record snow depths (Longhurst et al., 1952).

1910 A worse-than-average fire year (Clar, 1959).

1915-16 One of the biggest floods on Kern and Kings Rivers.

1919 Dry year at Tulare Lake. Severe forest fire year (Clar, 1959).

1921-22 Severe forest fire year (Clar, 1959).

1923-24 Low water on Kern River. Very serious fire years (Clar, 1959). Precipitation about one-half of normal.

1926 Bad fire year. Lumreau and Dennison fires.

1931-32 Heavy snow year. Snowed 12 feet at Camp Nelson in four days. Snow in Visalia (Doctor, 1959).

1933 Freak windstorm in August blew down a swath of of timber across Tule watershed on Dennison Mountain, and around Shake Camp, (Keagle, 1946).

APPENDIX B

1932-34 Drought. Low water on Kern River and historic dry years in Fresno.

1937-38 Flood on Kern River. Highest flood since 1867 on Kings River.

1940 Earthquake. Section line redwood on Camp Lena property fell.

1941 Windstorm on January 10 blew down great quantities of timber in Tule watershed.

INDEX OF PEOPLE
AND PLACES

See also pp. 154-8

Abernathy, Jake, 114, 119, 128
Ainsworth, Archie, 127
Ainsworth, Chet, 92
Akin, James, 52
Akin, Lola, Pl. 23
Allen, Howard A., 123
Alpine Meadows, 35
Amick, Harry, Pl. 12, 92, 94
Amick, John, family, 76, 92
Anderson, Marion, 74, 116
Atwell, H. Wallace, 22
Axe, Cromwell, 46, 47

Bace, Ed, ranch, 87, 128
Baker, Andy, family, Pl. 23, 104
Baker, P. Y., 33, 45, 66
Balch, A. C., 118-9
Balch Park, 3, 33, 51, Pl. 19, 101-2, 115, 118-20, 132, 144
Balch Park Road, 28, 89, 99, 110, 119-22
Ball, Eldon E., 115
Balwisha Indian tribe, 11, 28
Battle Mountain, Pl. 2, Pl. 3, 10, 15, 28
Bear Creek, 14, 20, 67, 86, 114
Bear Creek Ranch, 62, 114
Bear Creek Road, 28, 33, 51, 86, 89, 128
Beard, Bob, Pl. 25
Beeson, Wm. F., 91
Bennett, Ira B., 71, 129
Berry, Bill, ranch, 114
Blake, Frank "Kelly", Pl. 14, Pl. 23
Bland, Mack, 119
Blue Ridge, Pl. 2
Board of Forestry, 136-8
Bokninuwadi Indian tribe, 10
Boole, Frank A., 71, 84, 85, 143
Bostrum, Bob, 131
Bostwick, Henry, 31
Bowdlear, Lloyd, 121
Bowie, "Bull", Pl. 14

Boyd, Bill, 88
Bradley, Abel, 51
Brattin, Jack, 130-4, 138, 139
Breeding family, Pl. 14, Pl. 23, 88, 91
Brey, Hiram F., 79, 84, 94
Bright, Mandy, Pl. 14, Pl. 23
Brooks, Don, 125
Brough, Otis, 92
Brown, Albert, 92
Brown, Clinton T., 37, 42, 102, 155
Brown, Jay, 37, 42
Brown, Jess, Pl. 25
Brown, William W., 37
Brownie Meadow, 22, 30, 37, 91
Bruce, Jay C., 26
Burford, Will, Pl. 16
Burgess, Justin, 21, 87
Burns, Hazle, 91
Burns, Sen. Hugh M., 135-6
Burns, Matilda Watson, 91
Burro Creek, 24

California Division of Forestry, 121, 138
California Hot Springs, 7, 115
California Tree, 103
Calkins, John, 87
Camp, Pete, 76
Campbell, William, Pl. 14, 16, Pl. 23
Canty family, 70, 123
Carl, Luther, ranch, 49, 114
Carson, Kit, guide, 9
Carson, "Kit", teamster, 78, 92
Carter, Ed, 127
Centennial Stump, Pl. 5, 22, 50, 57-61, 99
Central California Redwood Company, 79, Pl. 11, 84, 94, 129
Cherbbonno family, Pl. 20, 87, 92, 95-6
Cherbbonno shake mill, 95
Chisel Mountain, 28, 34

INDEX OF PEOPLE AND PLACES

Churchill's, 50, 73, 86
Clark, Dalton, 121
Click, Martin, 28
Clicks Creek, 32
Coburn, Avon M., Front., 36, 51, 55-6, 68, 80, 85-8, 143
Coburn-Elster Mill, 87
Coburn's mills, 50, Pl. 8, Pl. 12, Pl. 18, Pl. 22, 63, 73, 74, 78, 84, 85-8, 94
Coburn, Samuel S., 36, 55
Collins, G. S., 67
Comstock, Smith, 69, 70, 89, 106, 143
Conlee, Frank, 91
Conlee Mill, 77, 91
Coso Mines, 31
Coso Mountains, 29
Coso Trail, 29
Cosper, E. T., 83
Courtner, Buster and Bounce, 132
Cowden, Henry, 33
Crabtree family, 24, 38, 87
Cramer, Eleanor, 46
Cramer, Jacob H., Front., 26, 46
Cramer, Lock, 94
Cron, Robert, 115
Crook family, 39, 46, 128
Cunningham, Della, 115-6
Cunningham, Frank P., Pl. 25, 115

Daunt, William G., 87
Davidson, 103
Davies, Ed W., 84
Davis, Samuel L., 128
Deer Creek, 7, 154
Demasters, David, 15
Dennison (the man), 24, 29
Dennison Peak, 2, 29, 40
Dennison Trail, 22, 28, 35, 89
Derby, Bill, Pl. 25, 115
Derby, Lt. George H., 7, 10
d'Heureuse, R., 24
Diaz, Ed, 31
Dillon, Alma, 34, 81, 85
Dillon, Delbert, 34, 74
Dillon, George, family, 34, 69, 91, 100, 126
Dillon, Henry, 34
Dillon, Ira B., 34
Dillon, Leander, 34
Dillon Mill, 28, 34, 36, 48, 51, Pl. 14, 77, 78, 80-2, 95, 97

Dillon Mill Road, 49, 83-5
Dillon, Nathan Patrick, Front., 34, 47, 48, 53-55, 69, 73, 143
Dillon's grist mill, 48, 53
Dillonwood, 36, Pl. 8, Pl. 11, 67, 69, 70, 83-5, Pl. 26, Pl. 27, 123-6, 127
Dillonwood Lumber Company, 83
Dillon Woods Corporation, 124-5
Doctor, Joe, 100
Doty, Andrew Jackson, Front., 90, 91, 98-101, 143
Doty, Charles B., 78, 87, 88, 101, 104
Doty, Elmer, 74, 87, 92, 94, 101
Doty, Jack, Pl. 12, 73, 74, 90, 91, 92, 94, 101
Doty, Sarah M., Front., 2, 69, 98-101
Doty's Resort, See Mountain Home Resort
Douglass, A. E., Pl. 5, Pl. 17
Doyle, Chester, Pl. 23, 69, 100
Doyle, John J., Front., 35, 51, 68, 69, 92, 101-2, 104, 118, 128, 129, 143
Doyle's Soda Springs (Doyles), 32, 35, Pl. 19, 101, 118, 121, 128
Drew, Arthur H., 133-9
Dubois, Coeurt, Pl. 25
Duckwall family, Pl. 23, 116
Dumont, Fred, Pl. 12, Pl. 14, Pl. 23
Duncan, Charlie, 85, 92
Duncan, Malvin, 30, 85, 94, 126, 127
Dunn, William, 51, 58, 60

Eckles, John G., 69
Eldridge, Mrs. E. E., 38
Elliott, Joseph P., 115, 130
Ellis, S. N., 46
Elson family, Pl. 23
Elster, Alonzo, 47
Elster, Chas. J., Pl. 5, 51, 52, 53, Pl. 14, 60, 78, 79, 86, 87, 91, 93-4
Elster, Grove, Pl. 9, Pl. 14, Pl. 23
Elster, Grandma (Rebecca), Pl. 23
Elster, Irvy, 26, 29, 47, 49, Pl. 23, 79
Elster Mill, Pl. 14, Pl. 16, 74, 77, 78, 80, 84, 93-5
Elster, Minnie, 94
Enterprise Lumber Company, See Enterprise Mill

INDEX OF PEOPLE AND PLACES

Enterprise Mill, Pl. 17, 64, 74, 75, 77, 78, 84, 85, 91-3
Estanislao, 9
Estudillo, Jose Maria, 7
Exeter, 28, 31

Farley, B. W., 2, 22, 29, 31
Farley, Minard H., 22
Ferry, Peter, 87
Flagg, Jim 46
Ford, J. P., 89
Fox, Kenneth, 115
Frasier "Grade", 51, 74, 75, 85, 88-91, 98, 110, 126
Frasier, L. B., 68, 88-91, 98, 143
Frasier Mill, Pl. 10, Pl. 15, 63, 69, 74, 84, 85, 88-91, 98, 103
Frasier Mill Campground, Pl. 24
Frazier Valley, 14, 55
Fremont, John C., 7
Futrell, Bud, Pl. 23

Gaffney, John and Ann, 63, 98, 128
Galena Cave, 63-4
Garces, Fray Francisco, 6
Garfield Redwood Grove, 40
Garner, Jake, 46, 63
George, Captain (Chief), 29
George, mountaineer, 119-20
George, Dr. Samuel S. G., 21, 31
Gibbons, D., 48
Gill, Charles, 48, 116, 128, 130
Gill, Clemmie, ranch, 48, 30
Gill, Fred, 126
Gill, John and Tod, 23
Gill, Levi, 39
Gill, Morris, 127
Gill, Ralph, 127
Gill, Will, 126, 127
Gilliam, S. M., 68, 69
Globe, 32
Goldsmith, C. B., 138
Greeley, Wm. B., 112-5
Green, Mrs. Emma Dillon, 48, 54
Green, Eugene, 80
Green, Fred, 80-2
Green Horn Gulch, 43
Greenewald, Louisa, 42, 70
Grider, William, 64
Griggs, Monroe, 74, 120
Griswold, Art O., 35, 52, 126-7
Grosse, Marion A., 75, 84, 85

Grouse Valley, 28
Guerne, G. E., 92

Haigh, George, 126
Hamar Valley, 132
Hammond, Mrs. John Hays, 118
Hammond, William H., 69
Hannum, Gen. Warren T., 136-8
Happy Camp, 49
Harbor Box Company, 124-5
Hardiman, Charlie, Pl. 14
Harding family, 115
Harper, Ben, Pl. 22, 128
Harrington, Robert, 47
Harris, Fred R., 84
Harris, Mal, 2, 124, 127, 132
Hart, Captain John L., 15
Hart, W. H. H., 89
Hatchet Peak, 46
Haughton, Edw. W., 51, 61-3, 86, 89, 143
Haughton's Cave, 57, 61-2
Hedrick, C. Fred, 121, 138
Hedrick's Mill, Pl. 28, 132
Hedrick Pond, 131
Hercules Tree, 99, Pl. 30, 127-8
Heston, Thomas H., 14, 43
Hickman, J. O., 83
Hilliard, Abraham, 14-16
Hilyard, Tom, 86, 88, 89
Hockett, John B., 32, 33
Hockett Meadows, 34, 38
Hockett Trail, 29, 33
Hodge, Allen, 86
Hodge, Eva Coburn, 87
Hodge, Warner I., 87
Hollow Log, Balch Park, Pl. 19, 51, 99, 102
Holman, Leslie J., 94
Holser family, 69, 101
Hoskins, Jesse, 62, 99, 126
Hossack, Belle Cramer, 12, 154
Hossack, John, Pl. 5, 38
Hossack Meadow, 3, Pl. 5, 37, 38, 115
House-That-Jack-Built, 133
Howe, Fred, 88
Howe shingle mill, 83, 95
Hubbs and Wetherbee Saw Mill, 47, 109
Hubbs, "Aut", 87, 94, 101
Hubbs, James Jr., 60
Hubbs, James R., 39, 47, 51, 53, 58, 142

INDEX OF PEOPLE AND PLACES

Hubbs, Milt, 91, 94
Hubbs, Mrs. Ola Doty, 21, 38, 63, 74, 85, 94, 98, 154
Hudson's Bay Company, 7
Hughes, lumber manufacturer, 84
Hume-Bennett property, 115, 128-39
Hume, George, 129-30, 133
Hume, Thomas R., 129-30, 143
Hyde, J. D., 68, 107

Indian "bathtubs", Pl. 2, 11, 12, 23
Indians, See various tribes
Inyo County, 22
Inyo Indians, 12, See Paiute Indians

Jack Flats (Jack-Ass Flats), 34, 47, 48
Jacobson Meadow, 32
James, Rass, 48, 87
Jarboe, John R., 70
Johnson, Amanda Dillon, 54
Johnson, George, Pl. 12, Pl. 14
Johnson, Walter, 134
Johnston, Anna Mills, 28
Jones, J. C., 123
Jones, Tom, 132
Jordan, Claud, 33
Jordan, John, 28, 31
Jordan, Captain John F., 33
Jordan Peak, 2, Pl. 2, 29, 40, 126
Jordan Trail, 29, 31, 34, 40, 51
Jordan, William, 121
Jordan, William F., 31
Joseph of the Nez Perce, 18
Judkins, T. C., 70
Junction Meadow, 67

Kains, Archibald, 71
Kaweah Delta, 27, 72
Kern Lakes, 23, 40
Kernville, 18
Kincaid, James A., Pl. 4, 48-9, 55, 61, 142
Kincaid, Maggie, 156
Kincaid Mill, 24, 34, 36, 48, 59, 86
Kincaid, Ray, 116
King, Clarence, 34, 38
Kinkade, Ellen Dillon, 54
Kirk family, Pl. 23, 127
Klingan, Bill, Pl. 25
Knowles, Frank, Pl. 3, 20, 26, 35, 50, 51, 63, 86, 92

Koheti (Coyohete) Indian tribe, 6
Kyle, J. W., 75, 92, 94

Lady Alice Tree, Pl. 19, 101
Lamb, Frank, 44
Lampe Lumber Company, 125
Lasure, Jim, 124-5
Lawson, John, 128
Lawson, Otis, 30
Lena, Camp, 35, 92, 99, Pl. 30, 126-8, 130
Levi, Jacob, 70
Lindley, A. C., 39
Lindsay, Clara A., 69
Lindsay, Lee, 92
Lindsay, Tipton, 107
Little Kern River, 28, 30, 33
Little, Ray, Pl. 30
Livingston, Captain, 16
Lloyd, R. C., 24
Lopez, Don Jose Jesus, 24, 41
Lumereau, C. M., 46, 51
Lumreau Mountain, 2, 46
Lyman, Bud, 125

Maggie Mountain, 30, Pl. 19, 67, 116
Manley, Ed, 58
Marshall, James, 54
Marshall, Nellie L., 54
Martin, Lyman, 33
Matthews, H. L., 31
Maynard, Carson, 125
Maxey, Mrs. Harriett, 155
McCoy, Mrs. Veda D., 53, 54, 85
McCutcheon, I., 51
McDonald, Earl, Pl. 7, Pl. 23, 76
McDonald, Eula, Pl. 23
McDonald, James E., 51, 52, Pl. 14, Pl. 23, 94, 115
McDonald, James, Jr., 126
McDonald, Joe, 52
McDonald, Pat H., Pl. 26, Pl. 27, 124-5
McFadgen, Allen, 70, 89, 91
McFadgen, Dan, 92
McFarland, Art, Pl. 16, Pl. 22
McFarland, Ed, Pl. 22
McFarland, Mrs. Martha, Pl. 22
McFarlane Toll Road Company, 33
McGee, Lizzie, 78
McKiearnan, John M., 51, Pl. 23, 58, 89, 102-4, 142

INDEX OF PEOPLE AND PLACES

McKiearnan, Oren, Pl. 23, 60
McKiernan, Pete, Pl. 14, Pl. 23, 85, 102
McKinney, Jim, 100
McNutt, Jack, 115
Meddick, Mrs. Edna, 85, 94
Metcalf, Cecil E., Pl. 30, 137-8
Michigan Trust Company, Pl. 29, 129, 138
Miller, Senator E. O., 109
Miller, Senator John F., 107
Millerton, 14, 40
Millinghausen family, Pl. 12, Pl. 23, 155
Mio, 10, 28, 74
Mineral King, 33
Mixter, Senator Frank W., 135, 136
Monache Indians, See Paiute
Monache Meadows, 32
Mono Indians, See Paiute
Moore, A. D., 70
Moore, A. R., photographer, Pl. 11, Pl. 23
Moore family, Pl. 23
Moore's Creek (Deer Creek), 7
Moraga, Gabriel, 6
Morton, James, 83
Moses Mountain, Pl. 3, 29, 32, 35, Pl. 19
Mount Whitney Power and Light Company, 118
Mountain Home Conservation Camp, 86
Mountain Home Mill, See Elster Mill
Mountain Home Resort, Pl. 16, Pl. 19, Pl. 21, Pl. 24, 21, 38, 69, 80, 90, 97-101, 115, 143
Mountain Home State Forest, vii, 44, 49, 67, 70, 91, 93, Pl. 26, Pl. 27, Pl. 28, 123, 128-139, 144
Mountain View, school, 59, 61
Moye, Lawrence, 119
Muir, John, 25, 39-40
Munoz, Fray Pedro, 6
Murdock, R. G., 127

Nash, T. J., 128
Native Sons of the Golden West, 133-5
Negus, Frank, family, 119, 126, 127
Nelson, Camp, 28, 37, 40, 160
Nelson, DeWitt, 136-8
Nelson, John, 126

Nelson's Fork of the Tule, 23
Nero Tree, Pl. 20, Pl. 23, Pl. 24, 103-4
Newhall, C. S., 109
Newport, Wm. J., 69, 89
Nichols, J. N., 123
Nofziger family, 75, 84, 143
Norris, Norman, 115-6, 130
Northrop, Clayton, 63, 114, 128
Norway, W. H., 45, 68
Nunes, John L., 64
Nunes, W. L., 63

Ogden, Peter Skeene, 7
Oldham, Alonzo, Pl. 14
Oldham, Mrs. Gertrude, 61, 99
Osborn, A. P., 51
Osborn, Clyde, 54, 74
Osborn, Perry, 74
Osborn, Thomas, 23
Owens Lake, 7, 18, 28, 29
Owens Valley, 11, 18, 28, 31, 33

Pacific Power and Light Company, 121
Packwood, Elisha, 14
Paiute Indians, 11, 12, 13, 28
Patterson, A. B., Pl. 25, 115
Peabody, Moses, 156
Peck's Cabin, 67, 128
Peck's Canyon, 115, 128
Pedigo, Tom, 74
Phariss, Tillman, 29
Phillips, Mrs. Irene, 79, 87, 98
Phillips, Lon, 79, 87, 91, 101
Pickard, Viro, 132; sawmill, 132
Pixley, Frank M., 128
Planchon family, 63, 114
Pleasant Valley, 34
Poindexter, Captain W. G., 15
Porterville, Pl. 5, 6, 72, 74, 75, 79, 103, 115
Powell, Alf and Monty, 64
Powers, Lucius, Jr., 135-6
Pratt, Mrs. Frances, 128
Pratt, Thomas, 128
Prescott, Fred, Pl. 14
Prescott, Rube, Pl. 14, 80, 94
Putnam, Porter, 25
Putnam, Will, 81

Ramsey, James, 23

INDEX OF PEOPLE AND PLACES

Rancheria, 10, 28, 30, 98
Rancheria Creek, 49, 73, 115
Rand and Haughton Saw Mill, 49, 52, 56, 73, 74, 86
Ratzlaff family, 63
Rauch, Claude, 130, 131, 138
Rauch property, 128
Rauch's mills, 124, 131, 133
Redfield, L. J., 100
Redington, P. G., 115
Reed, Forrest, 125
Richardson, Agatha, 157
Ridley, Alonzo, 16
Rogers, Dave, 88
Room Tree, See Hercules Tree
Root, Al, 125
Rose, William, 24
Ross, Andrew, 38
Roth, E. S., 123-4
Rutherford, Owen, 115
Rutherford, Wallace, Pl. 14, Pl. 23
Ryan, Frank, 30

San Cayetano, Rio (Deer Creek), 7
San Joaquin Light and Power Company, 118-9, 121
San Pedro, Rio, 6, 157
Sawed-off Tree, 102
Schofield, Wm., 136
Schutt, Harold, 84
"Sci-Con" Camp, 115, 157
Scruggs, Harry, ranch, Pl. 3, 16
Seamonds, Joshua, 61, 68, 74
Seamonds, Mary L., 89
Sequoia National Forest, 3, 88, 106-9, Pl. 25, 112-17, 130, 143
Sequoia National Park, 37, 44, 67, 106-8, 110, 121
Shake Camp, 22, 35, 127, 131
Sharp, Robert F., 38
Sheffler, W. J., 123
Sherman, C. E., 115
Shinn, Charles H., 112
Shirley, Elizabeth J., 69
Shuey, Ed, 87
Sickles, Ben T., 100
Sierra Forest Reserve, 42, 109, 112
Simpson, Clem, 79, 84, 85, 87, 92, 94
Slocum, Alvin H., 25, 98
Smith, Burk, 88
Smith, Charlie, 126
Smith, Hiram C., 70

Smith, Jedediah, 7, 13
Smith, O. K., 17; sawmill, 17
Smith, Pegleg, 13
Smith, W. T. F., 69
Snail Head, 9
Snider, Wesley W., 115-7, 120, 130
South Fork Meadows, 35
Southern California Edison Company, 121-2
Spangler, Charlie, 100
Spees, John, 85, 126
Springville, 26, 34, 39, 53, 59, 75, 84, 87, 115, 122
Stansfield, Jack, 77, 79, 81, 85, 94
Stathem, Paul W., 115
Steele, Ben, 124
Stevenson, Ray, 115
Stewart, Colonel George W., 106-8
Strathmore, 29, 30, 75
Street, James, 24
Street, Joseph, 23, 74
Stroud, Captain Ira, 15
Struble, Paul, 115
Success Dam, 10
Sullivan, W., 90
Summer Home, Pl. 4, Pl. 19, 97, 101-2, 118-22
Summit Lake, 2, 121
Sunset Point, Pl. 19
Sutch, Donald "Dude", 23, 116, 127, 130-3, 138
Sweet, S., 31
Sycamore Creek, 30, 39, 52

Tabor, photographer, 2, Pl. 22, 104
Talbot, Coleman, 69
Talbot, Courtney, 69, 70, 107
Talbot Meadows, 101
Talley family, 58, 87, 91
Talley, Fred, Pl. 23
Talley, Henry, Pl. 7, 58, 76
Taylor, Mervyn, 131; sawmill, 131
Tejon, Fort, 14, 16
Tejon Ranch, 41
Tharp, Hale, 25, 28
Thompson, Bill, 74
Thompson, G. W., 100
Tienken, A. B., 155
Tirrill, A. B., 84
Traeger, Henry, 64
Trout Meadows, 28, 30
Tubatulabal Indian tribe, 11

INDEX OF PEOPLE AND PLACES

Tulare Lake, 27, 159-60
Tule River Indian Reservation, 8
Tule River Lumber Company, 70, 84, 89, 106
Tule River Pinery Road, 49-50
Tuohy, John, 35, 38
Tuohy Meadows, 34, 35
Tuohy Sheep Trail, 35
Tyler, Clyde, 74, 87, 90

Uridge, Wm. G., 71, 84, 94, 129, 143
U. S. Forest Service, 35, 109, 112-7, 129-30, 136

Van Doorman, Neal, Pl. 22, 104-5
Van Doorman Tree, Pl. 22, 104-5
Van Leuven family, 53
Vasquez, Tiburcio, 30
Vincent, Earl, Pl. 23
Vincent, William Byron "Barney", Pl. 9, Pl. 10, 78, 81, 94
Visalia, 33, 34, 41, 52, 55, 59, 159-60
Vivian, W. F., 51
Vivian, Martin L., 51, 61

Wagy, P., 48
Wales, Rev. F. H., 128
Walker, Joseph R., 7
Walker Pass, 7, 29, 33, 113
Walton, Strother P., 133
Ward, V. L., 84
Warner, G. W., 32
Warren, Governor Earl, 137-8
Watkins, Oliver, 84
Watson, Wiley, 55
Webb, Jasper and Sam, 46
Weisenberger, Leo M., 72
Wells, Fred, 101, 156, 157
Western Pacific Lumber Company, 70
Wetherbee, Asa, 47
Wetherbee, Benjamin B., 47
Wheeler Expedition, 32
White, Bob, 87

White, Harrison, 42, 109
White, Huffman, 37, 42, 48, 109
Whitney, J. D., 38
Whitney, Mount, 20, 28, 38, 61
Wiesner, "Rat", 24
Wilcox, Marvin, 26
Wilcoxon, 34
Wilkinson, Harry, 115
Williams, Senator J. Howard, 138
Williams-Walker-Hooker cabins, 91
Wilson, Carr, Pl. 12, 94, 126
Wilson, Charles F., 49, 55
Wilson-Kincaid Mill, 49
Wimer's mine, 23
Wishbone Tree, Pl. 19, 102
Wishon, A. G., 121, 158
Wishon, Camp, 3, 20, 35, 101, 126
Wishon, Dave L., 121, 158
Wishon Fork of the Tule, 21, 23, 68, 69, 81, 102, 116, 121-2
Wishon Wagon Road, 35
Witt, Donald, 9, 120, 154, 157
Woods, "Doc", 123
Woods, Sandy, 74
World's Fair Big Tree, See Van Doorman Tree
Wortman, Dave, 52
Wortman, Earl, 52, 79, 84
Wray, George W., 39
Wray, George U., 128
Wright family, 81
Wynne, Sedman W., Pl. 25, 115

Yaudanchi Indian tribe, 9-19, 28
Yokod (Yokohl) Indian tribe, 28
Yokohl Valley, 17, 32, 48
Yokuts Indians, 5, 10, 11, 13
Yosemite Valley, 7, 62
Young, Ewing, 7
Young, J. W., Art, and Ed, 75, 83, 143

Zimmerman, Henry, 38

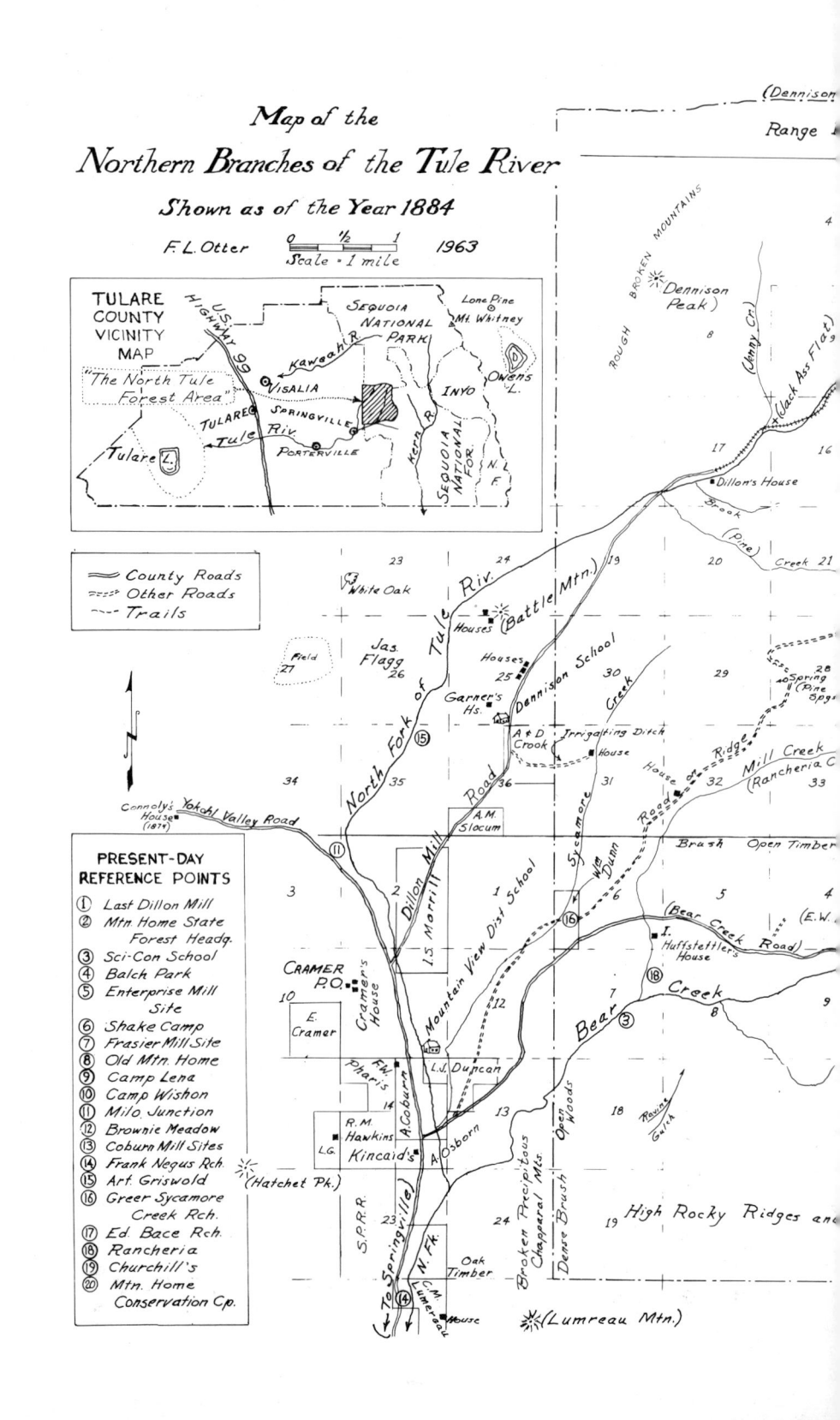

Map of the
Northern Branches of the Tule River
Shown as of the Year 1884

F. L. Otter   Scale = 1 mile   1963

TULARE COUNTY VICINITY MAP

"The North Tule Forest Area"

US Highway 99, Kaweah R., Visalia, Sequoia National Park, Lone Pine, Mt. Whitney, Owens L., Inyo, Tulare, Springville, Tule Riv., Porterville, Kern R., Sequoia National For., N.L.F., Tulare L.

— County Roads
---- Other Roads
--- Trails

PRESENT-DAY REFERENCE POINTS
1. Last Dillon Mill
2. Mtn. Home State Forest Headq.
3. Sci-Con School
4. Balch Park
5. Enterprise Mill Site
6. Shake Camp
7. Frasier Mill Site
8. Old Mtn. Home
9. Camp Lena
10. Camp Wishon
11. Milo Junction
12. Brownie Meadow
13. Coburn Mill Sites
14. Frank Negus Rch.
15. Art Griswold
16. Greer Sycamore Creek Rch.
17. Ed Bace Rch.
18. Rancheria
19. Churchill's
20. Mtn. Home Conservation Cp.

(Dennison Range)

Rough Broken Mountains, Dennison Peak, (Olney Cr.), (Black Ass Flat), Dillon's House, (Pine Creek), White Oak, North Fork of Tule Riv., Houses (Battle Mtn.), Jas. Flagg, Field, Houses, Dennison School, Creek, Spring (Pine Spgs.), Garner's Hs., A & D Crook, Irrigating Ditch, House, House, Ridge, Mill Creek (Rancheria C.), Connolly's House (1874), Yokohl Valley Road, Road, A.M. Slocum, Wm. Dunn, Brush, Open Timber, Dillon Mill Road, 1.5 Morrill, Mountain View Dist School, Sycamore, (Bear Creek Road) (E.W.), Huffstettler's House, Cramer P.O., Cramer's House, E. Cramer, L.J. Duncan, Bear Creek, Rough Gulch, F.W. Pharis, R.M. Hawkins, A. Coburn, A. Osborn, Open Woods, L.G. Kincaid's, (Hatchet Pk.), Broken Precipitous Chaparral Mts, Dense Brush, High Rocky Ridges, S.P.R.R., To Springville, N. Fk., Oak Timber, C.M. Lumreau, House, (Lumreau Mtn.)